SIMPLIFIED
STATISTICAL ANALYSIS
Handbook of Methods,
Examples and Tables

HARRY H. HOLSCHER, Ph.D.

Staff Scientist
Owens-Illinois, Inc.

Sponsoring Editor, Jerry Svec
Editor, *Ceramic Industry*

CAHNERS BOOKS, *Division of Cahners Publishing Co., Inc.*
221 Columbus Avenue, Boston, Massachusetts 02116

International Standard Book Number: 0-8436-0305-4
Library of Congress Catalog Card Number: 78-132672
Copyright © 1971 by Cahners Publishing Company, Inc.
Printed in the United States of America.
Halliday Lithograph Corporation, Hanover, Massachusetts, U.S.A.

DEDICATION

To W. L. Davis,
Associate, mathematician, and statistical expert,
who gave this manuscript the detailed editing
it required. His contributions are many.

PREFACE

Today, the engineer or research experimenter must get the most for his effort. Each day more companies are looking for higher technical efficiency. A daily usage of *Simplified Statistical Analysis: Handbook of Methods, Examples and Tables* will guide one in doing and analyzing just the required amount of experimentation. Your conclusions and recommendations will be based on that specified risk which balances the costs of experimentation versus the final output or product.

Furthermore, the examples give many numerical cases which allow quick desk usage — no computer wait or intermediate person to consult or convert — without a search for mathematical formulas, symbols, etc., which often make "projects" out of these techniques.

This engineering-type manual uses graphical types of solution wherever possible. Tabular layouts complete the "visual" techniques. Certain "factors," usually dependent upon the size of sample and the complexity of the analysis, are used where the visual procedures are not fruitful.

The main concept is comparison of sets of data for significant differences arising from proper experimentation and sample size. Limited introduction to quality control and control charts is required to complete the view of the original concept. But the full story of these latter items must be searched elsewhere.

The planning of experiments — that is, their "design" — is simply illustrated by a group of layout tables which almost explain themselves. Complete mathematical analysis of such planning is not necessary when the examples fill the gaps.

This book is compact, unified, and of easy access for quick and illustrative statistical analysis without the delay often necessary to review the mathematics. It is precise and complete enough for "desk-top" use for most cases faced by the practical engineer or research experimenter. It is also adaptable for the business-type analyst who is short on mathematical background.

H. H. HOLSCHER

CONTENTS

List of Figures   x

List of Tables   xii

List of Examples   xvi

1. INTRODUCTION   1

   1.1.   General Terminology   2
   1.2.   Terminology Related to Variables   2
   1.3.   Terminology Related to Attributes   4
   1.4.   An Outline of the Content of the Book   5

2. THE TECHNICAL DECISION   9

   2.1.   The Terminology of Technical Decisions   9
   2.2.   The Common Sense of Experimentation   10
   2.3.   The Human Parts of the Decisions   11

3. GENERALITIES OF EXPERIMENTS   13

   3.1.   Experience and Research   13
   3.2.   Sets of Conditions   14
   3.3.   Types of Experiments   15
   3.4.   Types of Results   16

4. EXPERIMENTAL DATA   19

   4.1.   A Single Measurement   19
   4.2.   Pairs of Values or Results   21
   4.3.   Groups of Data or Values   23
   4.4.   Rounding of Data   23
   4.5.   Presentation of Data   23

5. BIAS AND PREJUDICE   25

   5.1.   Definition   25
   5.2.   Examples   25
   5.3.   Conclusion on Bias   28

6. THE PRINCIPLE OF RANDOMIZATION   31

    6.1. Definition of Randomness   32
    6.2. Trend   32
    6.3. Statistical Control   32
    6.4. Effect of Change in Value of a Variable   33
    6.5. Summary   33
    6.6. Attainment of Randomness   33

7. PLANNING OF EXPERIMENTS   35

    7.1. Randomized Order with Replicates   36
    7.2. Replication   40
    7.3. Randomized Order without Replication   41
    7.4. Features of Various Plans of Experiments   47
    7.5. Some Larger Latin Squares   52
    7.6. Incomplete Experimental Blocks   54
    7.7. Other Incomplete Plans Illustrated   59

8. ANALYSIS OF EXPERIMENTAL DATA   63

    8.1. What do the Results Mean?   63
    8.2. Sigma   64
    8.3. Results with Normal Distribution   64
    8.4. Effect of Sample Size   66
    8.5. Graphical Procedures Using Probability Paper   68
    8.6. Mathematical Procedures for Sigma   77
    8.7. Sigma with Incomplete Data on any Specific Sample 84
    8.8. Limits of Uncertainty of an Observed Average   86
    8.9. Limits of Uncertainty of an Observed Sigma   92
    8.10. Uncertainty Versus Sample Size Plotted Graphically 95
    8.11. The Graphical Plot in Relation to Tolerance or
          Specification   97
    8.12. The Range (R) of Observations. Quick, Easy, Dirty
          Preliminary Analysis of Extensive Data   105

9. COMPARISON OF TWO SETS OF DATA FOR
    SIGNIFICANT DIFFERENCES   109

    9.1. The Simple Case   109
    9.2. The Graphical Procedure   109
    9.3. Determination of Whether there Are Sufficient Data
          for Establishment of Significant Differences   113
    9.4. The t Test   118
    9.5. Degrees of Freedom (df)   131

10. COMPARISON OF PRECISIONS OF TWO SETS OF
     MEASUREMENTS   The F Test   133

     10.1. Variance   133
     10.2. The Variance Ratio (the F Test)   136
     10.3. Analysis of Variance   145
     10.4. Short Cuts in Calculations for Variance and
            Analysis of Variance   152
     10.5. Interaction Studied by Analysis of Variance   154
     10.6. Sample Size for Estimation of Variance and Sigma   169

11. COMPARISON OF TWO SETS WITH DEFECTIVES
     FOR SIGNIFICANT DIFFERENCES. COMPARISON
     WITH ATTRIBUTES   171

     11.1. By Graphical Procedures   172
     11.2. Sample Size   173
     11.3. Average, Sigma, and Range for Average for Attributes 173
     11.4. Tests for Significant Differences for Attributes Using
            Large Sample   177
     11.5. The Chi-Square (χ^2)   181

12. THE QUALITY EVALUATION OF PRODUCTION LOTS   189

     12.1. Construction of Operating Characteristic Curves   190
     12.2. Interpretation of Operating Characteristic Curves   192
     12.3. Sequential Sampling   206
     12.4. The Sample Size (n) for Quality Evaluation   209
     12.5. Recommended Procedure for Determining Sample
            Size for Attributes for Quality Evaluation   213
     12.6. A Practical Example   216

REFERENCES   219

APPENDIX — Squares and Square Roots   221

INDEX   231

LIST OF FIGURES

8.1 (a) Normal Frequency Distribution for 10,000 Pieces   65
    (b) Percentage of Total Area Under a Normal Distribution
       Curve   65
8.2 Probability Paper Plot of Data from Figure 8.1.   67
8.3 Probability Plot of Tubing Diameter Versus Three Variables   72
8.4 Percent Sulfide Sulfur in One Amber Glass   73
8.5 Hydrostatic Pressure Strengths of 1,000 Bottles   74
8.6 Strength of Brick by Cell Plotting   75
8.7 Thickness of a Material Run in A.M. Versus P.M. of the
    Same Day   76
8.8 Plot of Data for Arithmetic Shortcut   78
8.9 Using Numerator Only for Probability Plot   80
8.10 Double Scaling Shows Multiplication of Average and Sigma   82
8.11 Plot of Factor K Values for Determination of Range for
    Averages   89
8.12 The Confidence Envelope for $n = 10$, $\bar{X} = 9.5$, and $\sigma = 2.0$ 96
8.13 The Confidence Envelope for $n = 100$, $\bar{X} = 9.5.$, and
    $\sigma = 2.0$ 98
8.14 Effect of Sigma on Rejections at Both Ends of the Curve   100
8.15 Case of Sigma Balance with Specification Limit for Low
    Loss   101
8.16 An Illustration of Wear on Loss of Defects Related to
    Starting Position   102
8.17 Relation of Tolerance Limit of Production and Rejections
    Given by an Inspection Device   104
9.1 The Case of No Overlap of 3σ Limits   110
9.2 Transverse Strengths of Two Lots of Glass   112
9.3 The Number of Tests Needed to Establish Significant
    Differences Between Two Sets of Data   115
9.4 Sample Size Required for Detection of Difference Between
    Averages   116
9.5 Glass Density Difference as a Result of Reannealing from
    no. 1 or no. 2 Strain Disks   122
9.6 Three Sets of Data on Strength of Glass   125

10.1 Transverse Strength of Glass Rods 143
10.2 Individual Data Probability Plot of a Glass Melting
Experiment 150
10.3 The Sums of Four Individuals Making up Groups of a
Glass Melting Experiment 151
10.4 Sample Size for Estimate of Standard Deviation to Within
Stated Percentage of True Value 168
11.1 Nomograph to Calculate (a) Significance for Attributes and
(b) Sample Size Required for Definite Significance 175
11.2 Charts for Confidence Limits for Averages for Limits 176
12.1 Cumulative Probability Curves from the Poisson
Exponential 191
12.2 Operating Characteristic Curves, with $n = 50$ and
$c = 0$ to 8 193
12.3 Operating Characteristic Curves, with $n = 100$ and
$c = 0$ to 5 195
12.4 Operating Characteristic Curves for Several Values of
n and c 200
12.5 Operating Characteristic Curves for a Lot of 10% Sampling
Plans, Each of Which Has a Producer's Risk of Rejection of
Lots Near 1.1% Defectives 202
12.6 Operating Characteristic Curves for a Set of Sampling
Plans Each of Which Has a Consumer's Risk of Acceptance
of a Lot of 5.3% Defectives 204
12.7 Characteristics of a Double Sampling Plan 205
12.8 Example of Sequential Sampling Based on Binomial
Distribution 208
12.9 Summation of the Poisson for Easy Estimate of c or More
Defectives 211
12.10 Sample Size Based on Observed Count, c, and as
Affected by Quality Level 215

LIST OF TABLES

5.1   Effect of Time of Pressing on Strength, Case I   28
5.2   Effect of Time of Pressing on Strength, Case II   28
7.1   Absurd Order with 4 Replicates   36
7.2   Absurd Order with 4 Replicates   37
7.3   Generalized Order Showing Blocks   37
7.4   Random Order Drawn from a Hat   38
7.5   Systematic Allocation of Treatment   38
7.6   Randomized Block   39
7.7   Rearrangement of Results in Order to Strike Averages   39
7.8   Spectrochemical Results   40
7.9   Tabulation for Computation of Table 7.8 Results   40
7.10 Data from Table 7.9 to Show Spread of Averages   40
7.11 Equipment as a Variable   42
7.12 Batch and Equipment as Variables   43
7.13 Averages of Molds and Cement Mixes   44
7.14 Cement Mixes and Molds — Three Factors, Three Levels   44
7.15 Reset of Table 7.14 Data   45
7.16 Latin Square, 2 × 2   46
7.17 Three Plans — 3 Variables, Each at 2 Levels   47
7.18 Classical Design   48
7.19 Factorial Design   48
7.20 Latin Square Design — Tabular Layout   50
7.21 Comparison of Designs — 3 Variables, Each at 2 Levels   51
7.22 Comparison of Designs — 3 Variables, Each at 2 Levels,
      Some Replication   51
7.23 Latin Square, 2 × 2 — 3 Variables, Each at 2 Levels   52
7.24 Latin Square 3 × 3 — 3 Variables, Each at 3 Levels   52
7.25 Tabular Layout (3 × 3) Latin Square   52
7.26 Another 3 × 3 Latin Square   53
7.27 Still Another 3 × 3 Latin Square   53
7.28 Latin Square, 4 × 4 — 3 Variables, Each at 4 Levels   53
7.29 Tabular Layout (4 × 4) Latin Square   53
7.30 Latin Squares, 5 × 5 — 3 Variables, Each at 5 Levels   53
7.31 Reducing the Trials Intelligently — A 2^6 Factorial   55

7.32 Balancing for Effects of A and D in a 2^6 Factorial   55
7.33 Reducing the Trials Intelligently in a 2^6 Factorial   56
7.34 Two Latin Squares with 3 Factors, Each at 2 Levels   57
7.35 Factorial Based on Table 7.34, Entering the Errors Only   57
7.36 Evaluating C   58
7.37 Comparison of B_1 and B_2 Errors   58
7.38 Blocking of Segments of Large (2^5) Factorial Experiment
     for Use of Four Lots of Raw Material   59
7.39 One Half of a 2^4 Plan   60
7.40 Eight Observations per Block in a 2^5 Factorial Plan   60
7.41 Incomplete Block Plan of 12 Experiments   60
7.42 Incomplete Block Plan of 20 Experiments   61
7.43 Incomplete Block Plan of 30 Experiments   61
8.1 Effect of Sample Size on Expected Average when Universe
     Average $= \bar{X}' = 200$ and Sigma Prime $= \sigma' = 20$ 66
8.2 Procedure for Graphic Plotting on Probability Paper   69
8.3 Calculating Sigma for 16 Successive Product Weights   79
8.4 Calculation of Sigma for Sulfide Sulfur in Amber Glass   81
8.5 Strength of Brick by Cells   83
8.6 Calculation of Sigma by Cells (Strength of Brick)   84
8.7 Sigma from Pairs of Data. Iron Oxide in Sand   85
8.8 Calculation of Sigma from Groups of Data. The Softening
     Point of Glass   87
8.9 Factors for K Values for Calculating Confidence Limits
     for Averages   91
8.10 t Values for Uncertainty of Average, Large Sample Size   93
8.11 Population Range, $\bar{X}'$, when $\bar{X}$ on 625 units $= 1000$ and
     $\sigma = 50$ 93
8.12 Factors for A_L and A_U Values for Calculating Confidence
     Limits for Sigma   94
8.13 False Acceptance and False Rejection   105
8.14 Using Range (R) Value to Estimate σ Versus Sample Size   106
9.1 Averages of Two Groups with No Significant Overlap   109
9.2 Strength of Two Groups of Glasses   111
9.3 Critical Values of t Based on Two Tails of Curve   121
9.4 Group Lots for Density of Glass   123
9.5 Time Interval, Seconds, for a Measured Process   123
9.6 Calculations for Example E9-11   124
9.7 Two Analysts' Results for Precision and Accuracy   126
9.8 Critical Values for t, 8 Degrees of Freedom   127
9.9 Pairs of Results — Durability of Glasses   129

9.10 Data for Paired Procedures, Lead Content by Two
Procedures   130

10.1 Data for Simple Analysis of Variance   135

10.2 Quality Evaluation Related to Shift, Date and Operator   137

10.3 The Variance Ratio F 139

10.4 Glass Rod Strengths   142

10.5 The F Test on Strength of Glass Compositions   144

10.6 Use of F Test to Compare Two Series of Analyses   145

10.7 Latin Square, 4×4, for a Glass Melting Experiment   147

10.8 A Glass Melting Experiment — Original Data   147

10.9 Analysis of Variance of a Glass Melting Experiment   148

10.10 Production Output (Units) Versus Facility and Worker   153

10.11 Reduction by Coding of Table 10.10: Subtracting 50 Units
from Each Value and then Dividing the Result by 10   153

10.12 Analysis of Variance. Production Output Versus Facility
and Worker   154

10.13 Original Data for Output Related to Shift and Male-Female
Crews   155

10.14 Analysis of Variance for Output with Interaction   157

10.15 The Original Data Layout for Interaction; 4 Variables,
3 Levels Each, 3 Replications   158

10.16 Summation of A Entries   159

10.17 Analysis of Variance   160

10.18 Interaction of A and C, Summating B_1, B_2, B_3, and
D_1, D_2, D_3 161

10.19 Interaction of C and D, Summating A_1, A_2, A_3, and
B_1, B_2, B_3 161

10.20 Interaction of A and D, Summating B_1, B_2, B_3, and
C_1, C_2, C_3 162

10.21 Interaction ADC, Summating B_1, B_2, B_3 162

10.22 Analysis of Variance for Study of Interaction   164

10.23 Two-Variable Interactions   166

11.1 Two Sets of Attributes   171

11.2 Ranges for Attributes from Figure 11.2   177

11.3 Comparison for Attributes — Example E11-6   177

11.4 Confidence Level Relation to Critical Values Useful for
Comparing Significant Differences of Attributes   179

11.5 Comparison for Attributes — Example E11-7   179

11.6 Comparison for Attributes — Example E11-8   180

11.7 Comparison for Attributes — Example E11-9   180

11.8 Data for Example E11-10   182

11.9 Values of Chi-Square   183

11.10 Observed Block of Data   184
11.11 Expected Block of Data   184
11.12 Block for χ^2 Formula   184
11.13 Data for Correction for Continuity   185
11.14 Two Processes with Three Lots of Raw Material   187
12.1   Poisson Series   190
12.2   Poisson Cumulative Probability for $n = 50$ and $n = 100$
       and Counts from 0 to 3   192
12.3   Quality Versus Sample Size with Count = 1 Defective   194
12.4   Sample Size and Average Quality   196
12.5   For Example E12-6 when $n = 50$, $c = 0$ to 8, $pn =$
       Variable   196
12.6   For Example E12-6 when $n = 100$, $c = 0$ to 5, $pn =$ Variable   196
12.7   Ninety-five Percent Confidence Intervals for Binomial
       Distribution   199
12.8   Data for Example E12-9   201
12.9   Effect of Lot Size When Sample is 10% of Lot   201
12.10 Effect of Sample Size on OCC for Similar Consumer's
       Risk of Acceptance   203
12.11 Effect of Sample Size on OCC for Similar Producer's
       Risk of Rejection   203
12.12 Double Sampling Plan, Example E12-12   203
12.13 Sequential Sampling Illustration   208
12.14 Calculation of Sample Size for Different Percent Attribute
       Quality Levels   212
12.15 Establishing Quality Tolerances   214
12.16 Sample (n) for Different Percent Defective Versus Count of
       Defectives   216

LIST OF EXAMPLES

E4-1. Micrometer Zero Shift   20
E4-2. Softening Point of Glass by Paired Values   22
E5-1. With Customer on Inspection Line   25
E5-2. Bias on Prepackaged Oranges   26
E5-3. Bias on Group Selection   26
E5-4. Psychological Bias   26
E5-5. Bias of Recording Data   26
E5-6. Bias Seeing Data Recorded   26
E5-7. Bias Selecting Products by Container Used   26
E5-8. Bias on Seeing Trademarks   27
E5-9. Bias on Established Limits   27
E5-10. Bias on Selecting Samples by Time   27
E5-11. Bias on Telephones versus Election Results   27
E5-12. Drift in Apparatus Bias   27
E5-13. Bias Due to Temperature Drift   27
E5-14. Bias Relating Nonrelatable Observations   27
E5-15. Bias on Reading Scales   27
E7-1. Spectrochemical Analyses Results   39
E7-2. Cement Molds, Mixing and Aging Time   43
E8-1. Sample Size versus Limits of Expected Average   66
E8-2. Glass Tubing Experiment   71
E8-3. Sulfide Sulfur in Amber Glasses   71
E8-4. Strength of 1000 Bottles   71
E8-5. Cumulative Grouping by Cells on Brick Strength   71
E8-6. Thickness of Material — A.M. and P.M.   71
E8-7. Short-cut to Calculation of Sigma   77
E8-8. Shifting of Zero Point (Scaling)   79
E8-9. Weight of Consecutive Products   79
E8-10. Dividing or Multiplying Sets of Data   81
E8-11. Sulfide Sulfur in Amber Glass   81
E8-12. Using Cells for Plotting Data   83
E8-13. Sets of Duplicate Runs of Iron Oxide in Sand   85
E8-14. Pairs of Data on Sodium Oxide in Glass   85
E8-15. Group of Data on Softening Point of Glasses   86

E8-16. Range of Average Indicating Poor Sampling   88
E8-17. Range of Average with a Low Sigma   90
E8-18. Time Interval in a Process   92
E8-19. Range of Sigma for Example E8-18   93
E8-20. Data for Plotting Confidence Envelopes, Sample Size = 10   95
E8-21. Data for Confidence Envelope, Sample Size = 100   97
E8-22. Dimensional Tolerance and Rejection   99
E8-23. Average and Sigma in Relation to Tolerance Limits,
       a Good Situation   99
E8-24. Shift of Average Giving Guidance to Selection Average
       within Tolerance Limits   99
E8-25. Confidence Envelopes versus Tolerance Limits   103
E8-26. Estimating Range of Losses (Rejections) from Confidence
       Envelopes   103
E8-27. Combining Quality of Product with Tolerance
       of an Inspection Device   103
E8-28. Tolerance of a Test Procedure   103
E8-29. The Range as a Measure of Sigma   106
E9-1. Range of Average and Significant Differences   109
E9-2. Data on Strength of Glass Rods for Plotting Confidence
       Envelopes   109
E9-3. Strength of Rods, Graphical Solution   113
E9-4. Out-of-Round for Significant Difference   113
E9-5. Strength of Rods, Calculating Sample Size   117
E9-6. Sample Size, New Product versus Old Product   117
E9-7. Sample Size, New Product versus Old Product   117
E9-8. Sample Size, New Product versus Old Product   118
E9-9. Example E9-7, Both Experimental Lots, Sample Size   118
E9-10. Glass Density Change Estimated to Probability Value   119
E9-11. Time Intervals for Four Conditions with Calculation of
       Significant Differences   123
E9-12. Three Sets of Glass Strength Data Compared for
       Significant Differences   124
E9-13. Group of Analytical Data Using Sum of Squares and
       Not Sigma   126
E9-14. Ten Pairs of Results on Durability by Sum of Squares   129
E9-15. Paired Results for Sigma   130
E10-1. Variance Ratio, Three Lots, Large Size Sample   138
E10-2. Five Sets of Glass Rod Strengths for F Value and
       Significance   142
E10-3. Two Series of Analyses for F Test   145
E10-4. Analysis of Variance, A Glass-Melting Experiment   146

E10-5.   Shortcut Analysis of Variance, Four Workers in Three
         Production Facilities   152
E10-6.   Quality of Output, Male and Female Crews, and Shifts   155
E10-7.   Analysis of Variance with Interaction, Four Variables,
         Three Levels, Three Replications   157
E10-8.   Sample Size for Sigma, Read Graphically   169
E11-1.   Graphical Reading of Significant Differences in Attributes   172
E11-2.   Graphical Reading of Significant Differences in Attributes   173
E11-3.   Graphical Reading of Significant Differences in Attributes   173
E11-4.   Graphical Reading of Significant Differences in Attributes   173
E11-5.   Range of Percent Defectives by Graph   173
E11-6.   Large Samples of Attributes for Significant Differences   177
E11-7.   Large Samples of Attributes for Significant Differences   179
E11-8.   Large Samples of Attributes for Significant Differences   180
E11-9.   Large Samples of Attributes for Significant Differences   180
E11-10. Chi-Square Test for Two Lots of Defectives   181
E11-11. Coin Tossing — Is Coin Biased?   185
E11-12. More on Coin Tossing   185
E11-13. Do the Two Coins Differ?   186
E11-14. Accidents versus Selected Months of Year   186
E11-15. More Data on Accidents   186
E11-16. Two Processes with Three Lots of Raw Material   187
E12-1.   Probability Values for Sample Lot of 50 from a
         Population of 2% Defective   190
E12-2.   OCC Interpretation for Sample Lot of 110 from a
         Population with 7% Defective   192
E12-3.   OCC Interpretation for $n = 30$ and One Defect Accepted   194
E12-4.   OCC Interpretation for $n = 50$ and One Defect Accepted   194
E12-5.   OCC Interpretation for $n = 100$ and One Defect Accepted   194
E12-6.   Comparison of OCC Interpretation for $n = 50$, $n = 100$,
         with $c = 0$ to 8 and 0 to 5 respectively   196
E12-7.   Range of Quality Determined by Table Usage   197
E12-8.   Reading Rejectable Quality Level (RQL) Based on
         Consumer's Risk of 0.10   197
E12-9.   Reading Acceptable Quality Level (AQL) Based on
         Producer's Risk of 0.97   201
E12-10. Four OCC Plans with 10% Sampling of Total Lot   201
E12-11. Four OCC Plans with a Narrow Range (Near 5.3%
         Quality) of Consumer's Risk of Acceptance   203
E12-12. A Double Sampling Plan   203
E12-13. Another Double Sampling Plan   205
E12-14. A Sequential Sampling Plan   206

E12-15. Sample Size by Table Readout   212
E12-16. Quality by Table Readout   212
E12-17. Quality and Sample Size on Basis of Five Units Rejected   212
E12-18. Sample Size Based on Count (c) for Rejection and
          Quality Level   214
E12-19. Sample Size Graphically for c (for Rejection) of 3 and 3%
          Quality Level   214
E12-20. Sample Size Graphically for c (for Rejection) of 5 and
          5% Quality Level   214

1

INTRODUCTION

"May I mow your
lawn, Sir? It's a
dollar if I try, and
three dollars if I finish."

"All is fair in love and war." However, even if you love your job, you may cause a war if your experimental planning is deficient. Let's take a general look.

Our technical community is composed of several groupings of rather diverse interests. For instance:

Fundamental research people are interested in *whether and why* a certain incident occurs.

Applied research people are interested in *how* a desired incident occurs.

Development people are interested in making the *incident occur once*.

Engineers are interested in *repetitive occurrence* of the incident.

Production and operating engineers will make the incident occur *repetitively at a profit*.

In all of these steps, data must be taken and analyzed. The material and process variables and control, the necessary equipment and layout, etc, are developed, revised, and refined as we proceed through the above groupings.

We must make order out of chaos — and do it economically. This book will help you to limit your data to what is required and also help you to interpret what your data mean.

1

1.1   GENERAL TERMINOLOGY

You must first recognize some new terminology. Even though restricting our mathematical usage, we will still use the recognized terminology.

A *unit* may be described as one of a number of similar items, objects, individuals, etc. A *sample* is composed of a group of units or portions of material, taken from a larger collection of units or quantity of material, which serves to provide information that can be used as a basis for judging the quality of the larger quantity. It may serve as a basis of action for acceptance or rejection of the larger quantity.

The *sample size* is the number of units in a sample. It may also be used in the sense of a number of observations in a sample.

A *universe* or *population* is the totality of the set of items, units, measurements, etc, real or conceptual, that is under consideration.

When we measure and record the numerical magnitude of a quality characteristic for each unit (we "read a scale"), we use the concept known as the *method of variables*.

When we note the presence or absence of some characteristic (property, character, quality, symbol, mark, trait, etc) in each unit, we use the *method of attributes*. A "go or no-go" gauging of a dimension is a prime illustration. A unit may possess several attributes.

Sometimes, we divide our observations into *subgroups* for convenience. They may be from the same or from different populations.

1.2   TERMINOLOGY RELATED TO VARIABLES

When a unit is *measured* for a quality characteristic, we get an *observation*, or better, an *observed value*. We designate a series of observed values as $X_1, X_2, X_3, \ldots X_n$, etc. Here n represents the number of units observed.

The arithmetic mean, or *average*, of a series of observed values gives us *X bar*, or $\overline{X}$. By equation, it is

$$\overline{X} = \frac{X_1 + X_2 + X_3 + \cdots + X_n}{n}$$

This $\overline{X}$ is often called the *sample average* and is the sum of the observed values in a sample divided by the number of observed values.

When our measurements get to where we think we possess the average of the *universe*, we designate this as X bar prime, or $\overline{X}'$, which equals the true or objective value of the average of that universe. If there are N items (values) in the universe, then

$$\overline{X}' = \frac{X_1 + X_2 + \cdots + X_N}{N}$$

The average of n sampled or observed values is the best estimator for $\bar{X}'$. The estimator $\bar{X}$ approaches $\bar{X}'$, with decreasing error, for large sample sizes (n approaches N).

The commonly known *standard deviation* is defined as the root-mean-square deviation from the universe average $\bar{X}'$. This quantity, herein designated σ', is

$$\sigma' = \left[\frac{(X_1 - \bar{X}')^2 + (X_2 - \bar{X}')^2 + \cdots + (X_N - \bar{X}')^2}{N}\right]^{\frac{1}{2}}$$

This value, σ', is a measure of the variability (spread) in experimental data. It is useful in data analysis.

The quantities $\bar{X}'$ and σ' are *theoretical concepts* and are rarely if ever obtainable in the laboratory. The number of items (possible samples) in the universe is usually very large. At best, the experimentalist will have data from only a small portion of the universe.

Now, if we want to take the average of a set of individual averages, we use X double bar or $\bar{\bar{X}}$ and get

$$\bar{\bar{X}} = \frac{\bar{X}_1 + \bar{X}_2 + \bar{X}_3 + \cdots + \bar{X}_n k}{n k}$$

If the samples are of unequal size, we may want to adjust to get a weighted average.

We have already stated that $\bar{X}'$ can be estimated from $\bar{X}$. The best estimate for σ' is the calculation of σ from the equation

$$\sigma = \left[\frac{(X_1 - \bar{X})^2 + (X_2 - \bar{X})^2 + \cdots + (X_n - \bar{X})^2}{n - 1}\right]^{\frac{1}{2}}$$

or for easier arithmetic

$$\sigma = \left[\frac{X_1^2 + X_2^2 + \cdots + X_n^2 - n\bar{X}^2}{n - 1}\right]^{\frac{1}{2}}$$

Here it is noted that σ differs from σ' in that $\bar{X}$ has been substituted for $\bar{X}'$, and ($n - 1$) for N. The reason for using ($n - 1$) instead of n is beyond the scope of this work (but see Section 9.5).

A number of estimates, say k, for σ may be made and averaged to estimate σ':

$$\text{estimate } (\sigma') = \left[\frac{(n_1 - 1)\sigma_1^2 + (n_2 - 1)\sigma_2^2 + \cdots + (n_k - 1)\sigma_k^2}{n_1 + n_2 + \cdots + n_k - k}\right]^{\frac{1}{2}}$$

If the samples are of equal size the above equation reduces to

$$\text{estimate } (\sigma') = \left[\frac{\sigma_1^2 + \sigma_2^2 + \cdots + \sigma_k^2}{k}\right]^{\frac{1}{2}}$$

Many texts use the symbol s for σ, and some erroneously use (n) as a division instead of $(n - 1)$. From a practical viewpoint, using (n) or $(n - 1)$ makes little difference even for moderate sample sizes. The error in using n is quite large for very small sample sizes. We recommend the usage of $(n - 1)$ as the divisor when calculating the standard deviation from data.

We will use the term sigma, σ, as a standard deviation, calculated from our sample.

1.3   TERMINOLOGY RELATED TO ATTRIBUTES

For contrast with the concept of variables, we earlier listed a general statement on attributes. We must expand this concept.

We introduce the term *event* as any kind of occurrence, failure, defect, etc, related to the consideration of a *trial*. A *trial* is a *single* instance of observing the presence or absence of an *event*. Thus, you see, in the field of studying attributes, we use the term *trial* as equivalent to the study of each unit, and the term *event* as descriptive of the attribute (comparable to the measurement of a variable). Here, though, the event is not measurable, but observed as present or absent.

The *relative frequency* (or proportion) is the number of events within all the trials under consideration. We designate this as small p or

$$p = \text{ratio of events to trials}$$

This case must be limited further to that where only one event can occur per trial.

If several events can occur per trial (say a bottle with three blisters present), then we may or may not wish to recognize them. Where a single event per trial occurs, we say the number of events in n trials is pn.

Where more than one event can occur per trial, then we call it small u and must consider our trial as a unit area of opportunity for occurrence of the event. Suppose we examine a 1 sq ft sheet of aluminum foil. We find a single pinhole. Here the trial is the examination of 1 sq ft for pinholes. We find 5 pinholes; the ratio of 5 to 1 gives u. We examine 5 sq ft (5 trials), and find 25 pinholes (events). The ratio is $25/5$ or 5 events per unit trial. The value u then is 5.

We like to designate the events in this case as small c to distinguish from pn above, which is used for the number of events where only one event can occur per trial.

We can now define a *defect* as a failure to meet a requirement imposed on a unit with respect to a single quality characteristic. It may be either a *variable*, or an *attribute*. A *defective* is a unit containing one or more defects with respect to the quality characteristic (p) under consideration. Furthermore, we define the term *sample fraction defective* as the ratio of the number of defective units in a sample to the total number of units in the sample. It is the number of defectives in a sample of n units divided by n. This we designate p; it is the same as the p used earlier.

When we multiply this p by the number of samples n, we get np or the number of defectives in a sample.

In our terminology, we say that c divided by n (c/n) is the sample defects per unit. We designate this as small u or

$$u = \frac{c}{n}$$

For all of our factors now described, we can also use averages as we did for variables. These are:

$$\bar{p} \qquad \overline{pn} \qquad \bar{u} \qquad \bar{c}$$

or

$$p' \qquad pn' \qquad u' \qquad c'$$

We can also use standard derivations of these symbols to get:

$$\sigma p \qquad \sigma pn \qquad \sigma u \qquad \sigma c$$

1.4. AN OUTLINE OF THE CONTENT OF THE BOOK

The prime consideration in all data gathering (experimentation) is to have everything under control. Here we include not only the raw material, the process and the equipment, but also the personal factors. The test of control is a repetition, or multiple repetition (known as replication), of the measurements.

After some definitions and generalities (Sections 2 and 3), we describe what a single measurement may or may not mean (Section 4.1). Pairs or groups of data are analyzed (Sections 4.2, 4.3).

Personal bias and prejudice are often difficult of control, and often occur in the simplest cases (Section 5). They are easily overlooked or even disregarded in cases of awareness. Weather variations, including humidity, air temperature, material temperature, equipment temperature, etc, etc, are very important.

An experimental variable, or factor, which is to be studied at several different levels should not be varied with time in an orderly increasing or decreasing level manner. The choice of order with time must be completely random (Sections 6, 7.1, 7.3).

Repeat measurements serve many purposes (Section 7.2).

In laying out experiments (design of experiments), we point to the usage of averaging or summating groups for major comparisons, even though such groups do not involve a single variable. Establishment of the experimental groups must be foreplanned to accommodate such a procedure; this will reduce the amount of experimental work (Sections 7.4, 7.5, 7.6).

The selection of these groups just mentioned may be such that errors existing in the small groups can be balanced out in our comparisons. These errors then have no effect on comparisons of levels of the main variables if the plan is good. Likewise, different raw materials may be used if such supply is limited; their effects can be nullified (Sections 7.6, 7.7).

In some cases incomplete plans of experiments are usable. This is particularly so if we can eliminate the variables having the lesser effects, or, vice versa, if we can discover the major variables by preliminary limited experimental planning (Sections 7.6 and 7.7). We also scan how to put the error where we want it (Section 7.6).

Very complex experimental situations can be initially simplified by breaking the plans down into several more simple cases, with each case limited to part of the variables (Section 7.6).

We use a graphical plan of solving average values and the standard deviation (Sections 8.3, 8.5). We find that graphical techniques for guidance of our sample size are useful (Sections 8.4, 8.10, 12). We also learn that certainty of results may be quite expensive and demands many experiments (Sections 8.8, 8.9). We must compromise.

It is wise to establish the *precision of test* for each specific case. This may be done by large series of similar nature, by replication many times, or by large series of paired values, etc. It should be done on a wide group of materials, preferably by several different operators, on widely ranging days or weeks, in summer and winter, etc, etc. Mostly we establish this precision over a period of experience (use) during many weeks or months. We discard those procedures which are disappointing; we adopt those that are fruitful. We describe an easily calculated ratio for this purpose (Sections 10, 11).

A brief discussion on the usage of "range" gives a little concept on "dirty" statistics (Section 8.12).

We must be sure that similar precision prevails for sets which are to be compared (Section 10). Averages don't mean much if precisions are all over the map (Sections 8.8, 8.9, 8.10). Comparison of results for significant differences takes several forms (Sections 9.1, 9.2, 9.4, 11).

A graphical method is shown to determine whether there are enough data to allow comparisons (Sections 9.3, 9.4, 12). We want sufficient data for our purpose, but more data may be costly and unnecessary.

We suggest several different methods for establishing confidences in average values. Significant differences, in variables (Section 9) or attributes (Section 11), are the prime points of comparing data.

The study of statistical precision can be very fruitful. The most detailed techniques — the Analysis of Variance — divide the total variation into useful and guiding concepts (Sections 10.3, 10.4, 10.5). Sample size is often larger than we think necessary (Section 10.6).

Different techniques may be desirable for studying attributes (defects), but the problems are of detail (Section 11), not of fundamental mathematics. Graphical procedure is useful (Section 11.1). More samples than commonly believed necessary are usually required. In this case any sample giving a zero, one, or two count of attributes is mostly meaningless. A count of at least three is required.

If our interest is in establishing a broad range of quality — and not comparison of two supposedly different qualities — the number of samples may be very large (Section 12.4). For a near perfect population, this may require a very big sample size, probably several hundred units (Section 12.5).

Some preliminary understanding is given for sampling and quality evaluation (Section 12), but only to stir your mind to go further if you have need.

2

THE TECHNICAL DECISION

There are five basic human errors:

errors of skill,
errors of ignorance,
errors of judgment,
errors of inattention,
errors of intentions.*

2.1 THE TERMINOLOGY OF TECHNICAL DECISIONS

Perhaps we start out with a fact. A *fact* is that which has existence, a thing done, a thing completed.

The fact is based on data. *Data* are material serving as a basis for discussion, inference, or argument; that which leads to an ideal or organized system of any sort.

The grouping of data leads to a result. A *result* is something obtained, achieved, brought about, etc, by calculation, investigation, or the like. It is an answer to a problem or knowledge gained by scientific inquiry.

We may have had a hunch. A *hunch* is a strong, intuitive impression that something will happen, or prove true.

Or an idea. An *idea* is a concept, representation, thought, image, notion, or design, which is a result of mental reflection.

Or a concept. A *concept* is an idea representing the meaning or essential attitudes of an action or thing; a mental image showing unity or universality of an action, or thing.

In gathering our data and facts, we must always be aware of the precision. The *precision* is the quality or state of exact determination of a set of data. It is the variation of data produced under set conditions; it is the smallest part that a system or device can distinguish.

*H. G. Butterfield, Quality Progress, 2, 9, (1969).

The precision must not be confused with the accuracy. *Accuracy* is the conformity to truth, or to an accepted value or standard.

Finally, we reach a conclusion. A *conclusion* is an inference; a reasoned judgment or an expression of one; an inductive generalization from a pattern of data.

We should use common sense. *Common sense* is a treasure-house of generalities that operate effectively most of the time in most circumstances.

And inductive reasoning. *Inductive reasoning* involves going from the specific to the general: the logic and theory of the methods and reasoning of empirical science; the bringing together of facts to arrive at a hypothesis or theory; or of a group of data to arrive at a consensus.

The consensus may be a hypothesis. A *hypothesis* is a proposition, condition, or principle which is assumed in order to draw out its logical consequences and to test its accord with factors which are known or may be determined; a preliminary estimate from a body of results or experiences.

Or a theory. A *theory* is an analysis of a set of facts in their ideal relationship one to another; a generalization with some uncertainty.

Or a principle. A *principle* is a generalization that has widespread application, but not to the universal extent of a law.

Or a theorem. A *theorem* is an established principle or law.

Or a law. A *law* is a representation of phenomena in an exact and universal fashion; it is expressed as rules and principles relating cause and effect in a selected subject, field, or area.

These all lead to a recommendation. A *recommendation* is a declaration of what one recommends.

Based on deductive reasoning. *Deductive reasoning* involves going from the general to the specific; a reasoning from a generalization, or law, to specific adaptations, or conclusions.

We reach what we recommend. To *recommend* is to commend, or bring forth explicitly, as meriting consideration, acceptance, adoption, election, or the like: to advise or counsel.

My recommendation to you is to go back and read this entire section very carefully again.

2.2   THE COMMON SENSE OF EXPERIMENTATION

In any experimental or other comparison, we must, at the outset, have an idea as to what differences between results are significant differences. Any comparisons may be so clear-cut as to be obvious, and conversely, they may be so obscure that they are not at all obvious.

The results are bound to show some scatter, and the scatter may offset the factors under study. *Statistical analysis* provides methods of treating data so that the maximum information can be obtained with a predetermined risk of drawing false conclusions. No method, including statistics, can draw conclusions from experimental data with zero risk of error.

It is this risk of error which one must evaluate before entering too far into the program. Do you wish to be right 19 times out of 20, or 99 times out of a hundred times? Many times you may fare well if you are right 60% of the time. But if a great capital investment depended upon the results, you may choose to be right 999 times out of 1000 times.

You will have some prior basis and knowledge on which to judge results. It probably will not be quantitative knowledge, but it should not be ignored. It is necessary to give careful thought to the size of the influences which it is important to detect, to the accuracy expected, to the significance level desired, and finally to the number of replications (repeat measurements) necessary to achieve these aims.

A variable may have less effect than the overall improvement desired, but a superposition of a number of such variables may have the effect desired.

The accuracy is selected as that which is economical to attain and necessary for the aims of the study.

The acceptable risk of decision should be set in advance and not after the results are in. It is far too easy to shrug off an unpopular result, after it is obtained, by readjusting the risk levels. All the intuitive and judgment decisions should be made beforehand and then used in setting the risk level.

The number of replications then comes up as a result of the factors entering into decision. The replication factors are very numerous and the decision here may come only gradually. In many cases, a preliminary statistical analysis will point toward the necessity or uselessness of additional replication. This involves an estimation of the relative differences in average results versus the statistical spreads or deviations from these averages.

2.3   THE HUMAN PARTS OF THE DECISIONS

Judgment is required to reach decisions, and to evaluate the proposed actions. The placement of a proper *value* upon experimental results leads to the conditional aspects of the *decision*. The decision often must reach the realm of conjecture — and, hopefully, also *prediction*. This is a kind of extrapolation only possible by *logic*.

Of course, the human aspects enter greatly in subjects such as experimental control, apparatus control, etc. The prediction must be sold by proper communication, which latter often involves psychological aspects such as emotion, feeling, prejudice, and conflicts with the opinions or ideas of other people.

In summary, there are three sources of error which are universal to almost all experimentation:

The variability of the material (sample) used.

The uncontrolled conditions of the testing.

The measurement errors of instruments and people.

So, think before you start taking data.

3

GENERALITIES OF EXPERIMENTS

People can be divided into three classes:

The few who make things happen;
The many who watch things happen;
The majority who have no idea what
has happened.

This book will help you get as much information as possible out of the least possible experimentation. You should find that interesting, since we are all lazy by nature.

In many types of experimentation, including all pilot types of operations, cost is of major concern. The research laboratory sells experimentation. Our product must be high in quality, low in cost, and completely competitive.

An *experiment* involves a *test* or *trial*. The work also has some connotation to the word *expert*. How expert are we? The purpose of this book is to set down in as simple a fashion as possible (and it is not always simple) some of the fundamentals of experimentation. The design of the experiment, the obtaining of data, the estimation of experimental precision, and the determination of significant differences are covered.

3.1 EXPERIENCE AND RESEARCH

Every day we have experiences. Some are pleasant. Some are not pleasant. Some are very old and repetitious in nature. Some are new and intriguing. Some we do not want to repeat. Some we want to remember. In any case, life is one experience after another. We try to control and develop the pleasant ones. We try to avoid those which do not advance us.

Briefly, therefore, no individual can avoid experiments. But some people, as a matter of personality and desire, enjoy new experiences and voyages into the unknown. When this unknown region involves science, we have the development of scientific inquisitiveness which all good experimentalists possess.

Frankly, if you do not enjoy probing the unknown, you are out of place in the research laboratory. Your job may then become a bore or a drudgery. It then may be more profitable for you, and for the company, to look for a job which better fits your personality and interests. You must be happy in your accomplishments. For research you must have a scientific interest and attitude.

3.2   SETS OF CONDITIONS

Each experience has a certain atmosphere or realm which contributes to it. These contributing conditions may not be recognized. Looking back, however, almost anyone can describe many conditions that were not evident at the time of the experience. In the field of human relations, the conditions are recognized, but are largely not definable. We all know that such things as "blue Monday," the most recent experience or argument on any matter, the lack of sleep, the weather, and numerous other things which affect human behavior, enter into a discussion. In this field the effects cannot generally be measured in terms of units or figures, which is one of the reasons why improvements in these types of relationships lag behind scientific improvements.

When one works with material things, rather than human relationships, the effects of all things entering the experiment can be at least partially controlled and measured. The ultimate aim, of course, is to control everything and thus have a *controlled experiment*. But no matter how carefully this is done, the one variable — the experimenter himself — always enters. When an honest effort is not made by the individual himself, all the efforts of control may be lost.

All science rests on the idea that similar events occur in similar circumstances. The similarities between events are usually restricted to a few features of particular interest, and then it is often found that "similar circumstances" can be identified by focusing attention on a rather small number of essential characteristics. The fixing of these essential conditions ensures the occurrence of a given event; we call these conditions *variables*.

The ultimate aim to develop an experiment under control may or may not be economically feasible. But where it is feasible, it should always be done. Particularly in pilot operation, we may spend a great deal of money developing a machine which exhibits a high degree of control on the process.

In experimental work we may often feel that an experiment is under control, but neglect some commonplace factor entering into it. As explained above, the individual may be an important factor, or the time of day, or humidity-temperature conditions may enter. Massive material may require a long time to come to the reaction temperature of its temperature of usage. Materials may come from an unheated warehouse. They may be contaminated by handling or condensation. There are many factors which may enter into or contribute to a loss of control of an experiment.

The notion that only a finite number of variables is sufficient to specify a given event is an idealization. Similar events of particular interest are indicated. As the class of similar events is more narrowly defined, the number of variables increases. The precision with which each has to be defined also increases. Variables fall into two categories: (1) controlled, which means they may or may not be measured; (2) uncontrolled, (a) those that are measured and (b) those that are unknown. Your own variable attitude may be very significant.

3.3   TYPES OF EXPERIMENTS

One experiment may be so costly that it can be done only once. Another may be done so easily that it can be repeated many times. Some experiments yield only one set of observations or data. Another may produce enough samples from which averages may be determined. None of these situations implies that the experiment is under control. They do, however, radically influence the plan of the equipment, the method of sampling, the taking of data, etc.

If an experiment is so costly that it can be carried through only once, then all details of control are so important that they need careful study and preliminary consideration. On an experiment which costs little in time and money, it may be economic at first to forego complete control to feel one's way. A group of pot shots set out to cover the extremes of conditions will probably lead to the best decision on the final controlled experiment. However, in all cases the procedure of measuring results at *extremes of conditions* may be considered. It may save much time, as certain variables may be found to be unimportant or to have little or no influence on the end result.

A repetition of an experiment serves several purposes. If a radically different result is obtained, then one is led to look for possible loss of control of the *constants* of the experiments. This may lead to consideration of variables which were not originally thought of, or which were felt to be under control, or felt to be unimportant. In any case, it opens your eyes. Too many experiments are reported which should be repeated. Time available and cost are the economic justification for or against repetition of experiments. But we should not let our own laziness be the reason for not having sufficient duplicate data.

If it is found by repeated experiments that the results are all as expected — not particularly the same — then we get some idea of the error in the experiment and the possible range of final results in production. A statistical treatment of the results may prove useful. Similar groups performed several days in a row, or with breaks in time and operator (experimenter), may be analyzed by *control chart* technique (Ref. 3) to give an idea of whether the experiments were actually under control.

In a group of routine determinations upon a single piece of apparatus — such as a group of glasses (with specific composition variables) in the softening point apparatus — the errors may be pretty well established by past experience. The *interrelationships* may prove useful in locating a single value that is out of line.

When all relevant variables are held constant, except the one under study, we have an ideal experiment but not necessarily the most fruitful experiment. Also, we have defined a situation which is not attainable in many cases. Some device then becomes necessary to help correct for the effect of other variables which may be changing in an unknown or uncontrolled manner. We may make use of *controls*, which are similar test specimens that are subjected to as nearly as possible the same treatment as the objects of the experiments, except for the change in the variable under study. In certain cases, the continuous use of controls may prevent an error where one may have a false confidence in one's ability to identify all the variables and keep them constant.

When the controls can also be called *standards*, an additional advantage is evident. Something against which comparisons can be made is called a standard if it is such that it can be reproduced by others or can serve as a common basis for cross-checking. The use of a control experiment is certainly desirable in many cases. The change in calibration of a thermocouple is often experienced. A control or standard would establish this quickly.

3.4 TYPES OF RESULTS

A result may be visual and described in a qualitative manner. A defect may be classified as qualitative or may be counted as unity in a quantitative fashion. Two defects in the same piece may be counted individually as a defective piece, or may double as two defects. Two blisters in a bottle may be counted as a blistered bottle, but may also be important because they fall at different locations. A blister and a stone in the same bottle may be counted as one defective bottle, but the importance of the difference is evident.

An article may be classified as "go" and "no-go" based on a measurable dimension such as the out-of-round limit. Here the measurement is made, and may not even be recorded. The degree of its variation from the accepted standard may receive no consideration whatsoever.

All of these types of results are classified as *attributes*, which means some qualitative or quantitative property of the article is being considered on a basis of "go" or "no-go" or on the basis of its presence or absence. The important part here is that the basis of acceptance is not distinctly a qualitative estimate. In fact, it may be quite quantitative and not an estimate at all. The attempts to put such ratings on a numerical basis may result in an average figure or number for quality. A completely satisfactory statistical analysis of such a situation may or may not be possible, but it may often prove advantageous. And again, averaging the defect situation may not prove at all satisfactory. In many cases, nothing better can be done.

This kind of estimation of quality, based on the presence or absence of certain undesirable (or desirable) attributes (or defects, if you please), has many uses in production where long continued runs are being made. In fact, it works much better under a repetitious condition than in a laboratory. It requires in general a great number of samples, or repeat observations, to establish a final definite conclusion. Of course, one can be happy where 100% of the pieces possess no defective attributes, and, conversely, one can reach a definite conclusion where all pieces are defective. The in-between conditions for establishment of, say, 95% passable quality may be very difficult to establish in a laboratory or a pilot operation.

We see a certain degree of quantitative estimation in the above discussion of attributes. When we get over to a full quantitative basis, we begin to attain a much improved condition. When we gauge a group of bottles for out-of-round, we secure a set of measurements much more subject to intelligent interpretation. We can get an average figure, establish a range of values, etc, and then decide whether our experiment means anything. We can change conditions, duplicate the measurements, and decide whether the change in conditions was important or contributory to the result. In short, by this method of *variables* we arrive at figures which allow us to say we are measuring the experiment in a quantitative manner. This sort of experiment is subject to much scientific evaluation and can be completely evaluated on a statistical basis.

All scientific personnel know the advisability of putting experiments on a quantitative basis and of making evaluations in terms of variables which are set down quantitatively. Our main point here is to stress this quantitative aspect and to open up the usage of the terms attributes and variables.

Another point which is quite important is that the usage of variables, measured in a quantitative manner, allows an average to be secured with a limited number of samples. At any particular five minutes in the operation of a bottle machine, we may take all the bottles, measure the out-of-round, and get an average value. However, one would hesitate to state the attribute quality, as regards blisters, on the same group of bottles. This statement applies regardless of the complete absence of blisters in the group. However, the presence of blisters in *all* samples could be regarded as a describable attribute.

4

EXPERIMENTAL DATA

"I am partial to Figures,
such as 37–27–35,
But it's the Facts
That should be Stacked."
(No replication necessary, — but fun.)
H. H. H.

4.1 A SINGLE MEASUREMENT

Suppose we have a chemical with a content of 5% of copper; that is, its true content of copper is 5.000%. This is the accurate copper content. We analyze it by a method which has been widely used in the laboratory and we obtain a value of 5.02% in a single determination. The analyst knows from his previous work on this method of analysis that his duplicate values seldom differ by more than 0.06 when the copper content is about 5% of the total sample. He could then expect a second determination to fall in the range 4.96% to 5.08%. But by experience he would expect the value to fall between 4.99% and 5.05% because most of the time the second determination falls within narrower limits. Supposedly, the average of a very large group of multiple determinations would fall at 5.00% — the accurate value of the unknown. But the average of the group is found to be 5.03%. This average is a precise determination by the method in question. The *precision* of the determinations falls within expectations, but the *accuracy* (difference from absolute truth) is not guaranteed. There appears to be a constant error, although precision is very good.

We are emphasizing that a single measurement by a *precision* technique may give an erroneous (inaccurate) value, but that this value will have some relation to subsequent values determined by the same technique. In the next section we will enlarge on this grouping of values as a function of the number of values determined.

Thus the concept of *precise* measurement has nothing in common with *accurate* measurement.

The accuracy of any determination or value is the conformity to truth, or to an accepted value or standard. The precision is an indication of the variations obtained upon multiple determinations using the same reproducible technique. Any constant error in the technique is repeated over and over.

EXAMPLE E4-1. In using a micrometer with a zero adjustment, a machinist may get a series of measurements on a 1.000″ round shaft which cluster at 0.985″. His range of measurements vary from 0.9835 to 0.9865 inches. He is immediately suspicious and checks the micrometer for its zero seat position, only to find that it is off 0.015 inch. He resets, and gets a new series of values of 0.9985 to 1.0015 inches. In each case, the precision is definite as $\pm$ 0.0015 inch, but the accuracy has been improved.

Thus let us conclude that any single value, or measurement, may have both accuracy and precision factors in it. Experience may guide us to an estimation of either. The precision of the measurement is the factor that is subject to some use of statistical technique. The accuracy may be estimated by several methods, the most common being to approach the problem by use of a different technique or procedure.

Comparative observations or measurements are all that are usually required. A comparison is the real object of the experiment and it may be costly to put the results on an *absolute basis*, that is, to express them in terms of universally accepted standards. If it is simple to put the results on an absolute basis, then do so by all means.

Inaccuracy must be recognized by the experimenter. It may quite often be obtained. It may not be important. Statistical techniques may often assist in locating inaccuracy, but such techniques are based primarily upon an analysis of precision. An inaccurate precise comparison may be useful, more so than an accurate imprecise experiment. The use of "imprecise" implies that a duplication may not give expected values, so the accurate imprecise determination was more or less accidental. Of course, the ultimate is the attainment of accurate values carrying precision.

Logically, we must be on the lookout when we make only a single determination by a technique we are not sure of. In the first place, it is very unlikely that the single value has accuracy. This follows because any one determination by an imprecise procedure is likely to have considerable error. (Please do not conclude from this that a group average of such determinations does not have such error.)

In the second place, under such conditions we are liable to let our thinking and common sense guide us to what I like to call a wishful result or value. If the first determination falls where we believe it should, we are sure it is correct. If it does not, we are likely to repeat the determination and if this satisfied our wishful idea, we accept it, and discard the first.

This procedure is also just as likely to happen when we do have a larger group of samples to select from. If the group as a whole averages, for instance, 0.025″ out-of-round, we select a sample to show someone; we may rather unconsciously pick one with an actual value of 0.010″ out-of-round. This develops from perhaps any one of several viewpoints. We may actually believe that we could, in a repetitive run, produce such an average condition. We have faith that we can better the situation as it was originally done. Or we may wish to impress someone. However, if this someone is wide awake, his first question, "Are they all as good as this?" may put us on the spot. This thinking obviously is not limited to a single sample or single determination. We may at times feel the urge to select 25 samples from a group of 100.

This *psychological bias* of the experimenter must be carefully watched. We all know that after we have a set of results we immediately look at them with the idea in mind of proving what we ourselves believe to be the case. While it may be quite evident that another conclusion is justified, we continue to juggle the data to fit our own preconceived notions. No human being is ever approximately free from these influences; only the naive or dishonest claim that their own objectivity is a sufficient safeguard. In the best experimental designs the person making comparisons, measurements, or records is kept ignorant of the identity of the subjects and controls. Yet, when we attempt to do this, we are accused of questioning the sincerity or honesty of the person involved. The conclusion is, "If you have a preconceived notion, it may be psychologically impossible for you to design the proper experiment or to analyze the data in an unbiased fashion."

How many times have you seen someone pick up a first sample, set it down, and pick up the second one to pass on to you?

The emphasis here is to be on the alert when you have only one value, a single measurement, or a single sample. Be sure that the expected precision applies. Be sure your experience justifies it. Do not be unconsciously affected by a wishful value.

Remember, don't lose your head over a fantastic figure!

4.2   PAIRS OF VALUES OR RESULTS

Everybody knows that two experiences are better than one; that is, you build up a better viewpoint which allows more confidence in the result. A repetition of a precise technique gives a second value which gives much more confidence to the first value obtained. Pairs of values rest on a much firmer basis than a single value. In the first place, a single value may have a gross error due to an oversight of a worker, such as misreading a scale, or transposing of figures in recording the result. Quite often the duplicate determination is made to avoid such gross errors, it already having been established that the technique, if followed satisfactorily, is totally reliable. Past experience provides the criterion for an acceptable agreement between duplicates.

Perhaps the use of duplicate measurements by a precision technique has been underemphasized in the ordinary work of an experimenter. While we cannot go into it yet, it is quite possible and very feasible to build up an estimate of the *standard deviation* of a technique by the use of duplicate measurements on a variety of samples.

EXAMPLE E4-2. Suppose that in the fiber softening point studies of glasses we have pairs of values on 20 different glasses. We take the difference in results for each pair, square it, summate the squares, and calculate the standard deviation as follows:

$$\sigma = \left[\frac{\text{summation of } (d^2)}{2 \text{ (number of pairs)}} \right]^{\frac{1}{2}}$$

Suppose for 20 different glasses we get summation of $(d^2) = 100$. Then

$$\sigma = \left[\frac{100}{(2)(20)} \right]^{\frac{1}{2}} = [2.5]^{\frac{1}{2}}$$
$$= 1.58°F$$

The meaning and usefulness of the standard deviation will be given later, but here let it suffice to say that two out of three measurements will deviate less than 1.58°F from an average determined on a group of multiple determinations.

The usefulness of pairs of data in establishing the precision of a technique may be compared to the usefulness of a very large group of similar measurements on a single material. In the illustration above, we used pairs of data on 20 glasses — we had a total of 40 determinations. In order to establish, with the same precision, the average deviation from a group of such measurements made on *one* glass, we would have to have 21 determinations on this single glass. In other words, n pairs of duplicates ($2 \times n$ determinations) furnish as much information as ($n + 1$) measurements all in one set. While more determinations are required using duplicates, the added information on the additional glasses may be the most economical way to establish the precision.

Again, as in single determinations, pairs of determinations give a greatly enlarged understanding, and this is again common sense.

4.3   GROUPS OF DATA OR VALUES

We do not need to emphasize further the importance of repetitive determinations. We all know that as the group becomes bigger, we can rely on it
in an increasing manner. We all also know that there is an economic limit
to repetitive determinations. It again becomes common sense to know
when to start and when to stop. (But we hope to show later that analysis of
data is also important.)

   This report is largely concerned with group data and so we need say no
more here.

4.4   ROUNDING OF DATA

All engineers have had the term *significant figures* drilled into them as far as
data taking and recording are concerned. Basically, we have been taught to
keep meaningful figures and discard all others. The old idea of rounding out
to the value of the precision of the measurement may prove to be very
erroneous. We know that we may distrust a scale on the sidewalk and not
believe it to be precise when we step on it. It may be good to three pounds or
so. If we stepped on it thirty times consecutively and only read it to the even
pounds — say 166, 164, 168, 162, etc — we would not end up with data good
enough to establish the actual precision of the scale. If we wish to establish
precision, we may consider it necessary to read to a degree much closer than
we think is required.

   The main point here is that rounding of data may suppress a large part
of the information originally available for determining the precision. The
extra decimal places that are of no consequence to the user are obviously
just the ones that do contain the information on the variation between determinations. It may often be advisable to retain in the data the terminal
decimal places which are often dropped. So the argument for rounding of
data can work both ways. It may be best to maintain the decimals and
round off only when the final report is being prepared.

4.5   PRESENTATION OF DATA

The field is wide open. The methods of presentation of data are manifold.
How much to present represents an ever-present problem. Representation as
a table, a graph, a chart, or as an equation must be considered. Even a
qualitative description of data is sometimes effective.

   The reader is the one who must get the picture. Briefly, then, the presentation should be geared to him. If his mind does not turn as a result of your
presentation, you have failed.

5

BIAS AND PREJUDICE

"Engineers and scientists seek facts,
Particularly those they are looking for."
H. H. H.

5.1 DEFINITION

We hardly need define bias or prejudice. One means you do not have an *indifferent mind* toward a subject and the other means you have an *opinion without knowledge*. We have mentioned the wishful result and the psychological bias. We have pointed toward the fact that we will go all out to prove a preconceived viewpoint. We will unconsciously credit or discredit data or weigh data in a manner to suit our whims. Somebody else usually manages to catch up to these situations. The biggest problem is to see that they do not enter into the planning of the experiment.

5.2 EXAMPLES

EXAMPLE E5-1.   You take a customer to a plant where his products are coming off the line. You each pick up a single item. You hope it is a good one which is representative of your production, or perhaps a little better than average. He picks up one which he secretly hopes will have a certain quality feature (note, I did not necessarily say defect, although that may be his state of mind) which he wishes to call to our attention. You lay your sample down; he immediately picks it up to see why. Here we have two states of mind — that of the producer and that of the consumer. Both may be biased or prejudiced — because they are trying to prove a preconceived idea.

EXAMPLE E5-2.   I go to the grocery store where the same oranges are available either in bulk for me to pick out or packaged in dozens. I pick out a loose dozen and put them in a bag rather than grab a package already bagged. I have a preconceived idea that I get better quality by picking out my own oranges. I may be prejudiced (without information) or actually have knowledge from prior experience that one of the dozen already bagged will probably be spoiled.

EXAMPLE E5-3.   A school carried out a test on relationship between the growth of children and the feeding of milk. One group was fed milk and the control group was not given milk. The results looked very good until it was discovered that the teachers had unconsciously selected an underweight group to receive the milk.

EXAMPLE E5-4.   In a test for the effectiveness of vitamins on colds, two groups were selected in an alternate fashion. Dummy pills made to look like the real vitamins were fed every other patient. The group getting the real vitamins reported an average reduction of 66% in number of colds contracted. But the results were no longer accepted as definite when the group getting the dummy pills reported a reduction of 62%. Such effects are purely psychological and must be carefully watched.

EXAMPLE E5-5.   In the recording of long lists of numbers or other simple data, it has been found that the mistakes in recording are usually more numerous in the direction personally favored by the recorder.

EXAMPLE E5-6.   In viewing a weak optical effect on the edge of visibility, an experimenter had an assistant operate a shutter in a random manner and keep the notebook. The assistant would either open or close the shutter, write down "o" or "c" and then record whether the experimenter did or did not see the light. The experimenter watched the writing, and knew whether "o" or "c" was written. The result was a preponderance of positive results — the open shutter resulted in seeing the faint light. When the experimenter was kept in ignorance of other symbols used for the shutter position, the correlation disappeared.

EXAMPLE E5-7.   A group of four pharmaceuticals of identical nature, but from different manufacturers, was sent to a hospital for comparison. Labels were removed from the bottles. It was found that the hospital favored those products which were in the bottles with caps which originated from that pharmaceutical house where they had always sent their orders. The package unconsciously was favored because of the shape of the bottle and cap.

EXAMPLE E5-8. We are asked to judge the color of competitive groups of products. Do we look for the trade marks? Or are we influenced psychologically by their presence?

EXAMPLE E5-9. We have an accepted lower limit for acceptance. Our production averages 5 units below this limit. Do we pass or reject? (But here some other problems of measurement procedure and tolerance of measurement will enter the picture. See Section 8.11 for example.)

EXAMPLE E5-10. We have sampled 25 products at 10:00 o'clock, a second group at 10:30, another at 11:00, 11:30, 1:30, and 2:00. The group at 10:00 is by far superior in quality. Which data do we use? Or do we make an intentional effort to get a group representative of the 6 hours' production? Do we try to find why the groups differ?

EXAMPLE E5-11. One of the most famous cases of the effect of a biased sample was the Literary Digest poll before the presidential election of 1936. Millions of ballots were mailed out and returned, with the result that the easy election of Landon over Roosevelt was confidently predicted. Unfortunately, the names of the mailing lists were mostly taken from telephone directories, and it turned out that a strong correlation existed between ownership of telephones (remember, 1936 was during economic depression) and a preference for Landon.

EXAMPLE E5-12. Duplicate analyses are run on a raw material. We have five such sets to run. Should they be run as follows?

Run	1	2	3	4	5	6	7	8	9	10
Material	A	A	B	B	C	C	D	D	E	E

Suppose a drift exists in an apparatus as it is used for 8 hours straight. A would check A, but the error in A versus E might be very different.

EXAMPLE E5-13. We have a plastic pressing machine and wish to determine the effect of time of pressing on the strength.

We start up and, in order, produce samples at 10 seconds, 20 seconds, 30 seconds, 40 seconds, 50 seconds, 60 seconds. These results are obtained! (See Table 5.1.)

EXAMPLE E5-14, *Bias in Conclusions*. "It has been conclusively demonstrated by hundreds of experiments that the beating of tom-toms will restore the sun after an eclipse."

How many times do we draw conclusions without trying the alternate?

EXAMPLE E5-15. In hydrostatic pressure testing of 1,000 bottles, it was decided that all test values would be read to five-pound intervals. After the data were in, it was found necessary to lump successive groups together as it was noted that many more bottles broke on the ten-pound interval than the five-pound interval. Apparently the testers had a tendency to read to tens much more frequently than to read to fives.

Need we illustrate further?

Table 5.1   Effect of Time of Pressing on Strength, Case I

Run	Time of Press in seconds	Strength in psi
1	10	2500
2	20	2750
3	30	3000
4	40	3050
5	50	3100
6	60	3150

Conclusion: Obvious — Time of press is important.
Experiment: Worthless — See Table 5.2 for repeat experiment.

Table 5.2   Effect of Time of Pressing on Strength, Case II

Run	Time of Press in seconds	Strength in psi
1	10	2500
2	50	2750
3	30	3000
4	60	3050
5	40	3100
6	20	3150

Conclusion: Machine Drift; mold temperature in this case. Time of press is unimportant!

5.3   CONCLUSION ON BIAS

It is probably impossible for anyone to free himself completely from preconceived prejudices, and in a certain sense this is not desirable. It is important to have some hypothesis in mind before making an observation; if this were not so, how would one know what to observe? On the other hand, it is equally important to arrange the conditions of observation so that the observer's bias will not distort the observation. This is far less easy than it sounds. The biggest problem is to see that bias does not enter into the planning of the experiment.

A scientific observer is never afraid to allow others to view the phenomena in which he is interested. He should welcome checks. When he gets defensive, others begin to wonder!

Another bias problem enters when we get observations which are out of line with expectations. The rejection of observations when once obtained is a difficult decision. If the reason for a "wild" result is obvious, the operator will know it and should discard the lot or experiment at the first known stage. It must be marked for rejection even though its final result comes out close to expectations.

The search for a specific reason for rejecting the observation is dangerous. Excuses, prejudice, and psychological bias will develop. It is important to look for causes of unusual results — as has often been shown when offshoots of major programs developed into important discoveries.

Negative results can never be discarded. Any excuses after a result has been obtained are suspect. Sometimes statistical analysis is helpful in these problems of rejection of results.

This subject now leads us into the concept of *randomization* (for experimenter, for time, for sampling, and for measurement).

6

THE PRINCIPLE OF RANDOMIZATION

"I love fools' experiments,
I am always making them."

We all know that results should not depend upon any of a group of personal or psychological factors. Such things as the operator, the day of the week, the time of the day, the weather, the lighting, etc, should be carefully considered. The order of the experiment and the plan of the experiment must be carefully studied to avoid false conclusions. We must look at all these, and perhaps many other things, as possible variables. They may later be proven to be of no import, but to assume at the outset that they are unimportant is pure folly.

In planning the work, we must adopt a randomness of these things. We must not let sample A always be worked on early in the morning, and sample Z late in the afternoon. We must not always let results be attained by stepwise variables in a progressive fashion, such as temperatures of 100°, 110°, 120°, 130°, 140°, 150°, etc. We must not always analyze our standard material the first thing on Monday morning. We must not always dictate our progress report immediately following a very good experiment.

The concept of randomness is based on the unpredictable. It can be likened to insurance where we may pay a heavy premium to protect ourselves against the unexpected. The premium may be worthwhile, as a "satisfaction" comes from the protection purchased.

Above all, we must adopt an *order of experiment* which tends to smooth out uncontrolled or unexpected variables.

6.1   DEFINITION OF RANDOMNESS

We know that we do not expect to get the same result each time. There are always sources of variability. However, the variations should form a *random sequence*. When the results are looked at in the order they were obtained, there should be no evident drift or trend. Such a trend shows that some important variable is not under control.

Randomness in a series of results is a very troublesome concept to define. The word itself has a connotation of chance, haphazardness, and the absence of definite aim, direction, rule, etc. In a general sort of way, it implies the absence of a pattern which could be used to predict what the next point is going to be.

The successive values may be very nearly the same in a very precise experiment, or they may scatter widely in an experiment of low precision.

6.2   TREND

A consistent *trend* in values indicates an experiment which is not under control. Unexpected and unpredicted variables enter, and the experiment must be carefully checked for these effects. At times the result is affected by the apparatus itself if it is affected by temperature, by materials which change on aging, by the effect of humidity and moisture, by changes in electrical characteristics of equipment and raw material, by contamination, leakage, viscosity of lubricants, and many other things.

Trends may often be broken by a *jump* in value. Here the variable may be easy to find, but it also may not be. Periodicity of results may require much search to discover the variable.

6.3   STATISTICAL CONTROL

When all consistent trends, jumps, etc in results have been brought down to a scatter which is random, the experiment or process no longer possesses variables which have large effects. All the major variables have been put under control. Many minor ones probably still exist, but some are positive, some are negative; individually they have small effects, but collectively they produce a scatter in the results. This scatter is random and the system is said to be in a state of *statistical control*.

A *sampling procedure (random)* and a *control chart* technique can now be used. The results are plotted as a function of time of production or output; the application of the *normal distribution law* is made, and limits are set within which the process remains under control.

Trends are not always indicative of a process out of control. With reference to control charts, a trend may indicate that a search for the cause is in order. Grant (Ref. 20) gives rules for deciding whether a change in process has occurred:

We may expect to find a cause where there are on the same side of the central control line, in successive points:

7 out of 7	or	10 out of 11	or	12 out of 14	or
	14 out of 17	or	16 out of 20		

There are many excellent works upon control chart techniques. This book does not cover them.

For a continuous process, the precision can now be estimated, and a determination made as to whether it is adequate or not. Furthermore, long runs may now give an idea of the ease of control of the process.

6.4   EFFECT OF CHANGE IN VALUE OF A VARIABLE

We are now ready to change a major variable and again put the experiment under control. The result of this experiment can now be compared with the first result, and conclusions can be safely drawn. We thus proceed through the range of values for a particular variable and plot its effect in a satisfactory manner. But in our changes in values of a variable we should again be guided by a *random* order. We should not progress from high to low temperatures progressive-wise, but experiment in a fashion that allows us, each time, to be on the lookout for drift. We should not work at 110°, 120°, 130°, 140°, 150°, etc, in order, but draw the order by lot from a random choice. Thus, the principle of randomness covers not only individual experiments where a single result is obtained, but also the process wherein continuous production is obtained.

6.5   SUMMARY

Randomness of experimental order is a fundamental procedure for good results. It accomplishes two things:

1. It prevents some overlooked effect from becoming identified with an experimental factor. (On this basis it is useful to judge the method.)

2. It ensures that any small overlooked effects are impartially distributed among the comparisons. (On this basis it is useful to judge the error.)

6.6   ATTAINMENT OF RANDOMNESS

We have stressed the importance of not letting some variable be progressively increased or decreased with time. The assignment of order of experiment can most easily be done by assigning each experiment a number at random, then drawing these numbers from a hat, and performing in the order drawn. This procedure can generally be used.

For more complicated experiments it may be desirable to use a set of random numbers in any one of several recommended fashions. References 7 and 10 give good illustrations.

7

PLANNING OF EXPERIMENTS

"The combined results of a
group of experiments should
yield a maximum amount of
information for the effort expended."

We need a good method of approach to experimentation. Our *plan* — our detailed program of approach, accomplishment, and analysis — involves an orderly arrangement of all parts of the program for maximum output with minimum effort. We need a *design* which fits our problem, our personnel, our process, etc, etc. Let us now define some terms.

Experimental design covers the arrangement in which an experiment is conducted and/or the choice of the combinations or factor levels which are studied; primarily a "beforehand planned grouping" for improved analysis and interpretation: an experimental pattern.

Replicate. One replicate results when all combinations of an experiment are completed. Further experiments give multiple replicate results. Lesser experiments give incomplete blocks.

Blocks are planned homogeneous groups where the experimental variabilities are expected to be small compared to the entire experimental layout; primarily, separations of experimental work into batches easily performed simultaneously (originated from block of ground).

An *incomplete block* results when the experimental work is reduced and a full replication is not accomplished.

Random order results when every individual (experiment) has an equal and independent chance of being chosen for a sample (as a first experiment).

A *randomized block design* results when the concept of blocking is used together with the concept of random order.

Factorial design covers all observations for all combinations of levels that can be formed from the different variables (originated from number of experiments making up a mathematical factorial).

The *classical plan* of experimentation involves changing only one variable at a time but with preselected levels of all other variables (originated as the classical thing to do).

A *balanced incomplete block design* assures that every combination of treatments appears equally often within an incomplete block.

A *Latin square design* results when the number of rows, columns, and levels of treatment requires triple blocking and equal balance so that each row, column, or level is represented by equal combinations; that is, the number of rows, columns, and treatments must be the same.

Interaction is the presence of a differential response due to shifts in level of one factor when these shifts are made at each of the various levels of one or more other factors.

These definitions and concepts will mean more to you later after you have seen the examples given in this chapter.

7.1   RANDOMIZED ORDER WITH REPLICATES

Suppose we make four treatments (A, B, C, D) each of which is run 4 times (1, 2, 3, 4). When a measurement is run only once, it is one replicate, etc; i.e., four runs mean four replicates. Each experiment takes one hour on each sample, so 16 hours are required for the total. A $\frac{1}{2}$ hour preparation or cleanup is required before each experiment or between experiments. One man is available to run these experiments. Thus, a day is required to run four experiments.

The First Design — An Absurdity.   We cannot distinguish the treatment variation if there happens to be a drift in apparatus during the day or a shift from day to day. This type of experiment (Table 7.1) should allow a comparison of the errors as a function of time of day.

Table 7.1   Absurd Order with 4 Replicates

Day of Week	Time of Day				Average
	9:00	*10:30*	*1:00*	*2:30*	
Monday	A1	A2	A3	A4	Mon, also av A
Tuesday	B1	B2	B3	B4	Tues, also av B
Wednesday	C1	C2	C3	C4	Wed, also av C
Thursday	D1	D2	D3	D4	Thurs, also av D
Average	9:00 Data	10:30 Data	1:00 Data	2:30 Data	

NOTES

frequency

lower yield point

→ stress

mean
calculated
stress

Y.P. Arbitrary

$$\text{Safety Factor} = \frac{Y.P. \text{ lower}}{\mu \text{ calc stress}}$$

Probabilistic Design asks how
much area overlap (shaded)
which would be failure
a small value, α.

Determine - Load distribution if you can

Prob. Des emphasize variation in physical
values, strengths, etc which is GOOD

Set Target of reliability .9 so it is
consistent with the consequences
.999999 for crew-return,

xcellent on static behaviour from
Haugens new book or drop from
a card

Mil handbook 5 B Raw Data at Battelle
Assumes normal Batch Catch

Next Year: 1977
DESIGN ENGINEERING
SHOW & CONFERENCE
McCORMICK PLACE, CHICAGO
MAY 9-12, 1977

Calendar of Events managed by Clapp & Poliak, Inc.

May 17-20, 1976
National Plant Engineering & Maintenance Show & Conference
Civic Center, Philadelphia

November 8-10, 1976
National Soft Drink Association's
International Soft Drink Industry Exposition
McCormick Place, Chicago

November 8-11, 1976
INFO 76 — Information Systems Exposition & Conference
McCormick Place, Chicago

November 9-11, 1976
International Pollution Engineering Exposition & Congress
Convention Center, Anaheim, California

May 9-12, 1977
Design Engineering Show & Conference
McCormick Place, Chicago

May 9-12, 1977
National Plant Engineering & Maintenance Show & Conference
McCormick Place, Chicago

July 19-21, 1977
Western Packaging Exposition
Convention Center, Anaheim, California

November 14-16, 1977
National Soft Drink Association's
International Soft Drink Industry Exposition
Convention Center, Anaheim, California

Overseas Events

September 27-October 1, 1976
2nd Japan Design Engineering Show & Conference
Harumi, Tokyo

September 27-October 1, 1976
2nd International Plant Engineering & Maintenance Exhibition & Conference
National Exhibition Centre, Birmingham, England

February 15-18, 1977
2nd PRO-IN (Product Innovation)
European Design Engineering Show & Congress
Messegelände, Düsseldorf, West Germany

Management
CLAPP & POLIAK, INC.
245 Park Avenue
New York, N.Y. 10017
Telephone: 212-661-8410
Telex: 12-6185
Cable: CLAPPOLIAK, NEW YORK

Consider the same experiment where all 16 trials could be made in one day (Table 7.2). Here a drift in apparatus or the progressive tiredness of the experimenter would upset comparisons of A averages with, say, D averages, or of the possibility of comparison of errors as a function of time. If the four values for D should have greater spread than those for A, we might not know just what to interpret. Results might not be independent of hour (time of day) differences.

Table 7.2   Absurd Order with 4 Replicates

			Time of Hour		
Hour of Day	0	15	30	45	Average
9:00	A1	A2	A3	A4	1st hour, also A
10:00	B1	B2	B3	B4	2nd hour, also B
11:00	C1	C2	C3	C4	3rd hour, also C
12:00	D1	D2	D3	D4	4th hour, also D

The term *block* refers to a group of results, say a particular operator or machine or time, planned in a natural fashion to be more alike than results from different blocks. In our case, these fall as segments of our tabular presentation — such as four replications. The generalized block appears in simplified form in Table 7.3. The arrows indicate the order of the experiments.

Table 7.3   Generalized Order Showing Blocks

Block	Treatment	Average
1	A → A → A → A →	A's
2	→ B → B → B → B →	B's
3	→ C → C → C → C →	C's
4	→ D → D → D → D	D's

Four treatments   A, B, C, D
Four replications   1, 2, 3, 4

Now change to random *order drawn from a hat* (Table 7.4). If the experiments are such that a trend might be expected from Block 1 to Block 4, we would be unable to detect it if the treatments had wide effects. Only Block 2 has all four treatments present. This is a result of chance order. If we are sure the block orders (order of experiment) mean nothing, then the random order has some drawbacks.

Table 7.4   Random Order Drawn from a Hat

Block	Treatment	Average
1	D → B → B → D →	x
2	→ C → D → A → B →	x
3	→ B → A → C → A →	x
4	→ C → C → D → A	x

We could rearrange the data, and average all four A values together, etc. In fact, this is what we always do. Then we have rather effectively eliminated time of day or order of experiment and have averages which are reliable.

Allocate each treatment to each block in a systematic manner (Table 7.5). Here the average of a column is independent of the block differences. The averages of the lines point out any effects of the block. If, however, the first experiment of each block was high, then the column averages may be incorrect. We thus have no comparison of treatments and no answer to our real problem.

Table 7.5   Systematic Allocation of Treatment

Block	Treatment	Average
1	A → B → C → D →	Block 1
2	→ A → B → C → D →	Block 2
3	→ A → B → C → D →	Block 3
4	→ A → B → C → D	Block 4
Average	A's B's C's D's	

Allocate each treatment to each block in a randomized fashion (Table 7.6). This is the *randomized block* with the order of experiments always being randomized. The true effect of *block differences* can be obtained from line averages. The true estimate of *error* is evident. It can be determined that error is largely a chance affair and independent of the order of experiment. When this is done, we are justified in rearranging the results and striking averages (Table 7.7). Here we are now not in order by time of experiment, but for averaging purposes only. The order is given in parentheses. A, B, C, D are treatments; 1, 2, 3, 4 indicate blocks.

All of these blocks have been illustrations of the study of one variable at four levels (A, B, C, D) or four treatments under similar conditions (such as four raw materials). We have illustrated the problem because they could not all be simultaneously studied (at the exact same period of time). The

Table 7.6 Randomized Block

Block	Treatment	Average
1	D → B → A → C →	ABCD
2	→ B → C → A → D →	ABCD
3	→ C → B → A → D →	ABCD
4	→ A → C → D → B	ABCD

Table 7.7 Rearrangement of Results in Order to Strike Averages

Block	Treatment A	B	C	D	Average
1	A1$_{(3)}$	B1$_{(2)}$	C1$_{(4)}$	D1$_{(1)}$	Block 1
2	A2$_{(7)}$	B2$_{(5)}$	C2$_{(6)}$	D2$_{(8)}$	Block 2
3	A3$_{(11)}$	B3$_{(10)}$	C3$_{(9)}$	D3$_{(12)}$	Block 3
4	A4$_{(13)}$	B4$_{(16)}$	C4$_{(14)}$	D4$_{(15)}$	Block 4
Average	A	B	C	D	

purpose was to set up the principle of randomization of tests so that time of test did not enter the picture. Furthermore, the replication required did result in an experiment where we were finally convinced that the experimental error was a chance affair. Thus, by randomization and replication, we feel sure that the error has been independently evaluated. Furthermore, we know the effect of our main variable — the four treatments, A, B, C, D — independent of time or other experimental facets of determination.

EXAMPLE E7-1, *An Example of Randomized Block.* The following is a comparison of five samples for nickel by a spectrochemical method. The five samples were run on each of five plates so that there are five randomized blocks. The order was *random* on each plate. For instance, on Plate I we first ran sample B, then A, E, D, and C in order. Note that each plate contains all five samples. However, each line does not contain all five samples, but by accident line 2 does show all five. The order and results appear in Table 7.8 (from Ref. 36).

Retabulate these data and obtain the results shown in Table 7.9. In this table we have averages for Plates I, II, III, IV, and V, as well as averages for samples A, B, C, D, and E. A common-sense spread of these averages shows the results given in Table 7.10.

We thus conclude that the variations in plates are much greater than in samples, or if you wish direct comparison of results, you must do the work on a single plate. The replication on five different plates has shown that samples A, B, C, D, and E are probably closely similar.

Table 7.8 Spectrochemical Results (from Ref. 36)*

Order	Plate I	Plate II	Plate III	Plate IV	Plate V
1	B 2.02	A 1.80	A 2.08	A 1.94	C 2.01
2	A 2.05	C 1.79	B 2.19	E 1.99	D 2.02
3	E 2.11	B 1.79	E 2.11	B 1.98	B 2.02
4	D 2.04	D 1.77	C 2.08	C 2.03	A 2.00
5	C 2.06	E 1.86	D 2.15	D 2.03	E 2.02

Table 7.9 Tabulation for Computation of Table 7.8 Results

Sample	Plate Number I	II	III	IV	V	Average
A	2.05	1.80	2.08	1.94	2.00	1.974
B	2.02	1.79	2.19	1.98	2.02	2.000
C	2.06	1.79	2.08	2.03	2.01	1.994
D	2.04	1.77	2.15	2.03	2.02	2.002
E	2.11	1.86	2.11	1.99	2.02	2.018
Average	2.056	1.802	2.122	1.994	2.014	1.9976

Table 7.10 Data from Table 7.9 to Show Spread of Averages

	For Plates	For Samples
	2.056	1.974
	1.802	2.000
	2.122	1.994
	1.994	2.002
	2.014	2.018
Average	1.9976	1.9976
Spread	−0.20	−0.02
	+0.12	+0.02

7.2 REPLICATION

Replication, according to the dictionary, is the act of reproducing. In statistics, one replicate means one complete set of all the experimental trials.

All too often the duplicate or replicate runs are conducted in parallel on the same day and under the same circumstances. This often results in highly satisfactory agreement among the replicate runs. If there are some temporary

*From W. J. Youden, *Statistical Methods for Chemists*, Wiley, 1951, by permission of John Wiley & Sons, Inc. Copyright 1951 by John Wiley & Sons, Inc.

uncontrolled circumstances, *both* runs respond together and the difference between these results gives no indication of the presence of this disturbance. Now the runs which involve varying the experimental factors are apt to be spread over the whole period occupied by the experimental program, and, quite apart from the deliberate alterations in the controlling factors, are subjected to a host of environmental variations which may also affect the result.

It is interesting that workers are aware that they may encounter trouble by making replicate runs of one experiment on Friday and comparing these with the runs of another experiment made on Monday. It is sensed that some circumstances may have existed on Monday which affected all results obtained that day. A difference between the averages for the two sets, it is clear, may be ascribed to this difference in circumstances as well as to a possible difference between the experiments. *This is exactly the trouble with duplicates run close together*. The question may be settled by repeating the first experiment on the second day (instead of running the second experiment). The persistence of a difference between the averages demonstrates that duplicates on separate days do not agree as well as duplicates run in parallel.

REPLICATION SERVES THREE PURPOSES:

1. To establish from the data an estimate of the error of the measurements.
2. To make possible some protection against faulty specimens and the inclusion of out-of-line measurements.
3. To gain the benefit of the increased precision of averages over that of individual measurements.

It is of no use to know one of the quantities very precisely if the other one is subject to large errors.

If both the method of determination and the selection of the sample affect the results, then some provision must be made for replicate measurements on the same sample by the same method.

7.3   RANDOMIZED ORDER WITHOUT REPLICATION

We have been talking about a plan of four experiments with four replications — a total of sixteen results. It is not often possible to repeat each experiment four times because of the demands of cost or of equipment available. If we have an independent measure of the experimental error and if we know an experiment is under control (there is no consistent drift), we may decide that doing the experiment once is enough.

By proper choice of variables, and when there is no interaction between variables, we may propose experimental designs which reduce the experimental work to a minimum. Here the term interaction implies that the effect of, say, variable *A* is the same regardless of the level of variable *B*.

The designs of *Section 7.1 are really designs of one variable studied at several levels. This is the simplest of all experiments.* Time of experiment or order of experiment must be removed as a variable. The experiment is under control.

When one studies the relationship and effects of more than one variable, the number of experiments is greatly increased. While a randomization of the order in which the experiments are run is possible and is highly desirable, a complete randomization using a number of replications may be impossible both in time, cost, and equipment available. If the equipment is not expensive, it may be desirable to have several experimental setups and thus reduce the cost of the experiments by performing in all sets of equipment simultaneously. Then the question of variability between sets of equipment may be a very real one. The equipment may become a variable.

EQUIPMENT AS A VARIABLE.   In Table 7.6 we used time as a variable or order of experiment as a variable. Now if we had four sets of equipment and could perform all 4 batches (A, B, C, D) simultaneously, we could set up a block as shown in Table 7.11, using the same order as in Table 7.6, Section 7.1.

While this is a randomized allocation of treatments to each block (horizontal lines), it is not randomized as to each equipment (vertical columns). Horizontal averages should be independent of any differences in equipment, but vertical averages cannot be compared as they are weighted by the variables of the treatments (A, B, C, D represent 4 batches). We have here 16 different experiments (4 batches and 4 equipments) and no replications. Each set of experiments with 4 equipments represents a block. Blocks necessarily represent 4 different experiments which cannot be carried on simultaneously. Thus, there is an order (by time) in this case. We cannot distinguish whether there is any difference in equipments.

BATCH AND EQUIPMENT AS VARIABLES.   Now we shall randomize the assignments of both Batch (A, B, C, D) and Equipment (1, 2, 3, 4) as shown in Table 7.12.

Table 7.11   Equipment as a Variable

Block	Equipment 1	Equipment 2	Equipment 3	Equipment 4	Average
1	D	B	A	C	ABCD
2	B	C	A	D	ABCD
3	C	B	A	D	ABCD
4	A	C	D	B	ABCD
Average	ABCD	BCBC	AAAD	CDDB	

Table 7.12 Batch and Equipment as Variables

Block	Equipment 1	Equipment 2	Equipment 3	Equipment 4	Average
1	A	B	C	D	ABCD
2	C	D	A	B	ABCD
3	B	A	D	C	ABCD
4	D	C	B	A	ABCD
Average	ABCD	ABCD	ABCD	ABCD	

Each line is an average for all 4 batches. *Each column* is an average for all 4 batches. Thus, errors due to different batches and different equipments are eliminated from the averages. There is no replication. While there may actually be differences in column averages, or in line averages, we are predominantly interested in *differences in batches*. Therefore, we average the 4 values for A, the 4 for B, etc, and ignore the averages of Table 7.12 except to study equipment or time variations. Note that we have eliminated equipment variations (if any exist) and time of experiment variations (if any exist). We really have used 16 experiments to get reliable averages for 4 treatments. However, none of the data are replicated except as you would consider the 4 A runs in different equipments as replicates.

AN EXAMPLE. The block design used in Table 7.12 has several unique features. Really, we have used:

4 equipments
4 batches divided equally to the 4 equipments
4 times of study, designated as Blocks 1, 2, 3, 4.

Because only one batch can be used at any specific time in a particular equipment, we cannot experiment with all three variables simultaneously. But suppose the third variable was temperature and not time. Now we have:

$$4 \times 4 \times 4 = 64 \text{ possible combinations}$$

of our three variables (batch, equipment and temperature). We obviously do not wish to run 64 experiments, but we must learn the effects of each variable.

EXAMPLE E7-2. To simplify this discussion, let us set up an experiment involving:

3 different molds (1, 2, 3)
3 different cement mixes (I, II, III)
3 different aging times (A, B, C)

in which we do not know for sure whether the molds are identical, whether the mixes are different, or whether the aging time is important.

Now there are 27 possible combinations of 3 variables, each at 3 levels ($3 \times 3 \times 3 = 27$). But we hope to get by with only part of this experimental drudgery. We set up a table which allows proper averages in all directions to be made, as Table 7.13 shows.

Table 7.13   Averages of Molds and Cement Mixes

		Mold		
Mix	*1*	*2*	*3*	*Average*
I	A	B	C	ABC
II	B	C	A	ABC
III	C	A	B	ABC
Average	ABC	ABC	ABC	

Putting in actual experimental values (Ref. 36), we get Table 7.14. This looks like a queer way of averaging values, but it does tell us right off that *the most unimportant factor* is the molds. Of course, by plain visual examination, we can tell that the most important factor is aging. Now we can reset the table using the mold as most unimportant and derive Table 7.15 (prefix indicates the mold number).

Table 7.14   Cement Mixes and Molds — Three Factors, Three Levels*

		Molds		
Mix	*1*	*2*	*3*	*Average (ABC)*
I	A 60	B 379	C 722	387
II	B 470	C 767	A 61	433
III	C 720	A 74	B 430	408
Average (ABC)	417	407	404	

We can conclude that the effect of aging is very significant because the A average also contains one result in each of the three molds (and for each of the three mixes), as does the B and C average. So mold effects are cancelled out or, rather, their effects are comparable in all three age averages. It is more efficient to determine whether the differences in mixes are significant. Note: The experimental plan allows an *analysis of variance* which will com-

*From W. J. Youden, *Statistical Methods for Chemists*, Wiley, 1951, by permission of John Wiley & Sons, Inc. Copyright 1951 by John Wiley & Sons, Inc.

Table 7.15 Reset of Table 7.14 Data

Mix	Aging Time			Average
	A	B	C	
I	1–60	2–379	3–722	387
II	3–61	1–470	2–767	433
III	2–74	3–430	1–720	408
Average	65	426	736	

pare the *error variance* with the *mix variance*, thus giving a significance or a nonsignificance to the effect of mixes. The original Reference 36 for these data gives these calculations for this case of no replication. It also covers the case where the data are replicated three times.

THE LATIN SQUARE. The foregoing material is an example of what is known as a Latin Square Design. At least three variables are necessary for the design, but one variable may be of minor importance. The name originated from the use of roman numerals, arabic numerals, and Latin letters for these variables.

There are two severe limitations on this Latin square design and the data layout.

1. There must be no interaction of variables; that is, the effect of one variable must be quantitatively similar whatever the magnitudes of the other variables may be. In this case aging must have similar effect whether we used molds (1, 2, 3) or mixes (I, II, III).
2. There must be an equal number of levels for all variables (in our case, 3, 3, 3).

These limitations may be very severe in many applications.

A Latin square may be defined as a desirable layout for the experimental work involving three or more variables which can be analyzed statistically to determine the effects of these variables. It involves no direct comparisons, and replicate runs are unnecessary if all variables are under control. In the case of four or more variables, the experimental error is arrived at statistically.

Our purpose here is not to use the statistical basis, but to show the use of Latin squares in a simpler, but not as precise, procedure.

The *simplest Latin square* is made up of the three variables at each of two levels. Assume these factors to be: C = concentration; T = temperature; and P = pressure. Subscripts C_1 and C_2 represent the standard level and the trial level respectively. Table 7.16 gives the 2×2 Latin Square (known as 2×2 because it involves 4, or 2×2, experimental results).

Table 7.16 Latin Square, 2×2

	C_1	C_2
T_1	P_1 (1)	P_2 (7)
T_2	P_2 (5)	P_1 (6)

Here we run four combinations or four experiments as follows:

Experiment 1 $C_1 \, T_1 \, P_1$ Experiment 2 $C_1 \, T_2 \, P_2$
Experiment 3 $C_2 \, T_1 \, P_2$ Experiment 4 $C_2 \, T_2 \, P_1$

Note that there is no direct comparison for any one variable. That is, we have violated the rule to which we are so accustomed, i.e., that of having only one variable changed at a time. However, the method is sound mathematically.

We analyze the data by striking averages. We use 1, 5, 6, 7, for a reason that will be apparent soon, to represent our results from Table 7.16.

$$\text{Effect of pressure} = \frac{(1 + 6) - (5 + 7)}{2}$$

This is the difference between P_1 and P_2 values.

$$\text{Effect of concentration} = \frac{(1 + 5) - (6 + 7)}{2}$$

This is the difference between C_1 and C_2 values.

$$\text{Effect of temperature} = \frac{(1 + 7) - (5 + 6)}{2}$$

This is the difference between T_1 and T_2 values. Or

For pressure — compare major diagonals
For temperature — compare major lines
For concentration — compare major columns

Thus, *while direct comparisons involve averages of different things, we are making these comparisons by averaging two results*, and not by a single result.

Furthermore, we have determined the effect of three variables with only four experiments. An important consideration is that, if we care to analyze the data in a completely statistical sense, we have all that is necessary for the process. The analysis of variance may be made and the error can be assessed.

7.4 FEATURES OF VARIOUS PLANS OF EXPERIMENTS

There are three basic plans of experiments when a number of variables are encountered. Obviously, the complete plan is one where all possible combinations of variables are worked together, each combination giving an experiment. This requires a very large number of experiments, a procedure which is often unnecessary. It is known as the *factorial plan*.

The classical plan involves changing only one variable at a time, but with preselected levels of other variables — not all levels of other variables. This program may result in some useful conclusions, which may be limited by the question of interaction between the variables. Interaction in a classical plan may lead to very sad situations.

The Latin square plan is a selected series of combinations which is statistically sound.

Let us illustrate further by assuming three variables, C, T, and P, each at a standard sub_1 and sub_2 level. The total number of experiments for one trial at each level is as given in Column 1 of Table 7.17. In this table there has been no replication. This we will get into somewhat later. It is also now evident why we designated experiments of Table 7.16 as 1, 5, 6, and 7.

Table 7.17 Three Plans — 3 Variables, Each at 2 Levels

Experiment Number	Combination	Classical	Latin Square	Factorial
1	$C_1\ T_1\ P_1$	1	1	1
2	$C_2\ T_1\ P_1$	2	—	2
3	$C_1\ T_2\ P_1$	3	—	3
4	$C_1\ T_1\ P_2$	4	—	4
5	$C_1\ T_2\ P_2$	—	5	5
6	$C_2\ T_2\ P_1$	—	6	6
7	$C_2\ T_1\ P_2$	—	7	7
8	$C_2\ T_2\ P_2$	—	—	8

THE CLASSICAL DESIGN. *The classical design* involves the use of two levels of each variable with a single selected level of the other variables. This is classical since it is an early accepted procedure. Taking the classical combinations of Table 7.17:

Compare Experiments 1 and 2 to get the effect of concentration, but note that we get this at only one level of T and P.

Compare Experiments 1 and 3 to get the effect of temperature, but note only at one level of C and P.

Compare Experiments 1 and 4 to get the effect of pressure, but note only at one level of C and T.

In each case, the levels of the unchanged variables must be set to the more useful level and without further experimentation your selection may not prove the best. Furthermore, there are no data on which to base:

a. The experimental error
b. Whether the differences are significant
c. Whether there are interactions between variables; that is, does a change of temperature radically affect the result when pressure is being studied?

The tabular layout for the classical design is given in Table 7.18.

Table 7.18   Classical Design

	C_1		C_2	
	P_1	P_2	P_1	P_2
T_1	(1)	(4)	(2)	
T_2	(3)			

THE FACTORIAL DESIGN.   The factorial design given in Table 7.17 includes all possible experimental combinations. If we have three variables, each at two levels, we thus get $2^3 = 2 \times 2 \times 2 = 8$ possible combinations. This is the reason for calling it factorial. Thus, the tabular layout shows no blanks, as given in Table 7.19.

Table 7.19   Factorial Design

	C_1		C_2	
	P_1	P_2	P_1	P_2
T_1	(1)	(4)	(2)	(7)
T_2	(3)	(5)	(6)	(8)

The effect of concentration results from:

comparison of (1) versus (2)
comparison of (4) versus (7)
comparison of (3) versus (6)
comparison of (5) versus (8)   and perhaps averaging all.

The effect of pressure results from:

comparison of (1) versus (4)
comparison of (3) versus (5)

comparison of (2) versus (7)
comparison of (6) versus (8)   and perhaps averaging all.

The effect of temperature results from:

comparison of (1) versus (3)
comparison of (4) versus (5)
comparison of (2) versus (6)
comparison of (7) versus (8)   and perhaps averaging all.

Furthermore, striking averages for major effects, we get:

Average effect of concentration:

$$\frac{(1) + (4) + (3) + (5)}{4} - \frac{(2) + (7) + (6) + (8)}{4}$$

Average effect of pressure:

$$\frac{(1) + (3) + (2) + (6)}{4} - \frac{(4) + (5) + (7) + (8)}{4}$$

Average effect of temperature:

$$\frac{(1) + (4) + (2) + (7)}{4} - \frac{(3) + (5) + (6) + (8)}{4}$$

Basically, therefore, *we average four results* to get our overall effects of any one variable. But we still do not know if a repetition of the experiments would give the same effects; that is, we have no idea of experimental error. We can get this by replication or by analysis of variance.

This plan (as differing from the classical) allows us to get a measure of interaction. Take concentration, as shown above:

(1) vs (2)   (4) vs (7)   (3) vs (6)   (5) vs (8)

all give effect of concentration. Now, if the difference between (1) and (2) is only 5 units, and the difference between (5) and (8) is 25 units, we know that concentration is more important when temperature and/or the pressure is higher. Thus, an interaction exists for the variables. In the classical design we have no experimental data for (5) and (8) and so the interaction would go unnoticed.

There are certain homogeneous subgroupings which can be best selected from factorial designs, if the experimental work must be reduced (see Section 7.6).

THE LATIN SQUARE DESIGN. The Latin square design represented in Tables 7.15 to 7.17 results from assigning *at random*

each treatment within a row   and   each treatment within a column

Thus, P_1 and P_2 are found in each row *and* each column of Table 7.16. All treatments must appear in each row *and* each column. The preferred tabular layout is given in Table 7.20. Note that this is $\frac{1}{2}$ of a 2^3 factorial plan — 4 out of 8 experiments.

Table 7.20 Latin Square Design — Tabular Layout

		C_1		C_2
	P_1	P_2	P_1	P_2
T_1	(1)			(7)
T_2		(5)	(6)	

The effect of any one single variable cannot be determined on the basis of any set pairs of direct comparisons for such pairs do not exist. However, the effects of variables may be determined by striking averages which can be determined as follows:

The effect of concentration results from comparison of average of (1) and (5) to average of (6) and (7), or

$$\frac{(1 + 5)}{2} - \frac{(6 + 7)}{2}$$

The effect of temperature is, similarly,

$$\frac{(1 + 7)}{2} - \frac{(5 + 6)}{2}$$

The effect of pressure is

$$\frac{(1 + 6)}{2} - \frac{(5 + 7)}{2}$$

We compare the

major columns	for effect of C
major lines	for effect of T
but the major diagonals	for effect of P

All these three totals are directly comparable without adjustment.

SUMMARY OF DESIGNS AND THEIR CHARACTERISTICS.   We have considered the three designs with three variables each at two levels. Summarizing, we arrive at Table 7.21. Note that "no" in the column for Interaction Measured indicates that, if interactions exist, the estimated effects are meaningless.

Table 7.21   Comparison of Designs — 3 Variables, Each at 2 Levels

Type of Design	Runs Required	No. Tests Averaged	Experimental Error Measured Directly	Interaction Measured
Classical	4	1	No	No
Latin Square	4	2	No	No
Factorial	8	4	No	Yes

Now in order to measure experimental error, we must repeat or replicate measurements for the classical and the Latin square designs. A complete analysis of variance for the factorial design will yield an estimate of experimental error, but replication of a portion of the design may be easier and cheaper in many cases. It depends on the complexity and cost of the experiments. The availability of computers may alter this above view. However, the most useful replication may be on the Latin square design, if one is sure that interaction does not occur. Table 7.22 gives a comparison of these designs.

Table 7.22   Comparison of Designs — 3 Variables, Each at 2 Levels, Some Replication

Type of Design	Runs Required	No. of Tests Averaged	Experimental Errors Measured		Interaction Measured Directly
			Directly	Variance Analysis	
Classical replicated	8	2	Yes	Yes	No
Latin square replicated	8	4	Yes	Yes	No
Factorial	8	4	No	Yes	Yes
Factorial replicated	16	8	Yes	Yes	Yes

Here we see that the 2 × 2 Latin square is $\frac{1}{2}$ the factorial in runs required (also compare in Table 7.17). *This situation is still more favorable for the larger Latin square designs.* We note later in Section 7.5 that a 3 × 3 square is only $\frac{1}{3}$ of a factorial; that a 4 × 4 square is only $\frac{1}{4}$ of the factorial, etc. We also note that replication is unnecessary in 3 × 3 or larger Latin squares in order for them to be used statistically for estimation of error. Furthermore, the larger the Latin square, the more precise is the estimation of error.

7.5   SOME LARGER LATIN SQUARES

We wish to point out again that the Latin square gives data which, if desirable, can be analyzed by the analysis of variance. It is the minimum block plan for any set of variables which can be so analyzed. Because of this property and of other uses shown in Section 7.6, we show some illustrations of the larger plans (Tables 7.23, 7.24, 7.26, 7.27, 7.28, 7.30). Table 7.25 is a tabular form of Table 7.24; Table 7.29 is a tabular form of Table 7.28. Note that in a 3×3 Latin square, it takes three such Latin squares to make up the factorial. Also in a 4×4 Latin square, four such plans make up the factorial.

In Table 7.30 the subscripts get too complicated. Note the regularity of the letters in anglewise fashion. This gives the concept for laying out larger Latin squares. See Reference 26 for illustrations.

A basic feature of a Latin square is that the number of rows and columns must be equal. Totals of rows, columns, and the third variable are directly comparable without adjustment.

Table 7.23   Latin Square, 2×2 — 3 Variables, Each at 2 Levels

	C_1	C_2
T_1	P_1	P_2
T_2	P_2	P_1

Table 7.24   Latin Square 3×3 — 3 Variables, Each at 3 Levels

	C_1	C_2	C_3
T_1	P_1	P_2	P_3
T_2	P_2	P_3	P_1
T_3	P_3	P_1	P_2

Table 7.25   Tabular Layout (3×3) Latin Square

	C_1			C_2			C_3		
	P_1	P_2	P_3	P_1	P_2	P_3	P_1	P_2	P_3
T_1	(1)				(5)				(9)
T_2		(2)				(6)	(7)		
T_3			(3)	(4)				(8)	

Table 7.26   Another 3×3 Latin Square

	C_1	C_2	C_3
T_1	P_2	P_3	P_1
T_2	P_3	P_1	P_2
T_3	P_1	P_2	P_3

Table 7.27   Still Another 3×3 Latin Square

	C_1	C_2	C_3
T_1	P_3	P_1	P_2
T_2	P_1	P_2	P_3
T_3	P_2	P_3	P_1

Table 7.28   Latin Square, 4×4 — 3 Variables, Each at 4 Levels

	I	*II*	*III*	*IV*
1	A	B	C	D
2	B	A	D	C
3	C	D	B	A
4	D	C	A	B

Table 7.29   Tabular Layout (4×4) Latin Square

	I				II				III				IV			
	A	*B*	*C*	*D*	*A*	*B*	*C*	*D*	*A*	*B*	*C*	*D*	*A*	*B*	*C*	*D*
1	×					×					×					×
2		×			×							×			×	
3			×				×			×			×			
4				×				×			×			×		

Table 7.30   Latin Squares, 5×5 — 3 Variables, Each at 5 Levels

	I	*II*	*III*	*IV*	*V*
1	A	B	C	D	E
2	B	C	D	E	A
3	C	D	E	A	B
4	D	E	A	B	C
5	E	A	B	C	D

7.6   INCOMPLETE EXPERIMENTAL BLOCKS

We have implied that statistical analysis of an incomplete set of data or block design is not possible. However, this is not necessarily true if the incomplete block is properly laid out. Our main interest here is to show incomplete blocks which are useful where there are many variables and complete experimentation is too costly or even impossible. Furthermore, by proper grouping of portions of blocks into simultaneously run experiments where exact control is impossible, or where there is a limited amount of raw material (and one must lose control by usage of several lots of raw material), we may eliminate the effects of such loss of control or of limited material. *Averaging is so set as to be self-balancing.* We will now develop these themes. *The scheme of selection is based on the Latin square concept and this is perhaps its most useful purpose.*

REDUCING THE EXPERIMENTAL TRIALS.   Let us take a common problem. We develop that there are six different factors which *may* prove important. In order to study this many factors, we must study each of them at two levels. This means $2^6 = 2 \times 2 \times 2 \times 2 \times 2 \times 2 = 64$ experiments for a complete factorial set up. We normally would not do this. So, we must pick combinations from these 64 possible experiments and do it intelligently.

The tabular layout for this case is given in Table 7.31 which shows a 2^6 factorial with 6 factors, each at 2 levels. Suppose we approach the problem first of determining whether the variables A and D are important ones. We wish data to average to make such a determination possible and also such that the data can be used later in a properly designed plan. Thus the data must be properly balanced as to choice of the remaining variables.

We must have an equal number of trials at A_1 and A_2, as well as D_1 and D_2, if we are to average different groupings of data. Likewise, if the data are to be used later for effects of the other variables, B, C, E, and F, then these groupings must also be balanced. First, in order to balance A_1 and A_2, we must have at least one trial under the A_1 portion as well as one under the A_2 portion of the table. In order to balance D_1 and D_2, we must have one trial under each. This gives four trials to start, but we cannot balance B_1 and B_2 unless there are equal numbers of trials under their columns. The same argument holds for balancing E, as well as C and F. One might easily conclude that we must fill out the table to attain these balances, but such is not the case. Table 7.32 shows a balanced incomplete plan which is obtained by four different Latin squares (one for each corner of the table).

What we have done here is to select alternate values constant for each of the four Latin square areas. Remember, the smallest Latin square is one with 3 variables each at 2 levels. So we reduce to 3 variables for each corner; these variables are different for each corner. In each case we select some other variables as constant and balance these others for the whole table.

Table 7.31   Reducing the Trials Intelligently — A 2^6 Factorial

			A1				A2			
			B1		B2		B1		B2	
			C1	C2	C1	C2	C1	C2	C1	C2
D1	E1	F1								
		F2								
	E2	F1								
		F2								
D2	E1	F1								
		F2								
	E2	F1								
		F2								

Table 7.32   Balancing for Effects of A and D in a 2^6 Factorial

			A1				A2			
			B1		B2		B1		B2	
			C1	C2	C1	C2	C1	C2	C1	C2
D1	E1	F1	1			1				
		F2					1			1
	E2	F1		1	1					
		F2						1	1	
D2	E1	F1					1			1
		F2	1			1				
	E2	F1						1	1	
		F2		1	1					

The result of this procedure is:

Upper left corner	A1 constant, D1 constant, F1 constant
Lower left corner	A1 constant, D2 constant, F2 constant
Upper right corner	A2 constant, D1 constant, F2 constant
Lower right corner	A2 constant, D2 constant, F1 constant

Since it takes four trials for the smallest Latin square, we allow four trials for each case, and on this basis of six variables it is obvious that we arrive at sixteen trials.

If you study Table 7.32 intelligently, you will see balance for all variables. In other words, even though we started out for balancing A and D, we now find we could have

$$8 \text{ for } F_1 \qquad 8 \text{ for } D_1 \qquad 8 \text{ for } A_1$$
$$8 \text{ for } F_2 \qquad 8 \text{ for } D_2 \qquad 8 \text{ for } A_2$$

and so on. Now we proceed to average the 8 for F_1 and compare the 8 for F_2 so we get the effect of F. Likewise,

$$A_1 \text{ vs } A_2 \qquad B_1 \text{ vs } B_2 \qquad C_1 \text{ vs } C_2 \qquad D_1 \text{ vs } D_2 \qquad E_1 \text{ vs } E_2 \qquad F_1 \text{ vs } F_2$$

In each case averages for 8 values are used. Obviously it is a precise comparison.

Note that no experiment has been replicated. *If you trust single values or the average of two values or four values, you may make studies of interaction.*

This is thus the Latin square for the six variables, each at two levels. From it we determine the two or three, etc, most unimportant variables and proceed further from there.

Suppose we find that A and D are both unimportant at the higher levels. Then we confine further experimentation, if we feel it necessary, to the upper left quarter of Table 7.32. (Depending on our selection, we could have continued further experimentation in any segment of Table 7.32 or even selected new levels for certain variables.) The further experiments could come as shown in Table 7.33.

Table 7.33 Reducing the Trials Intelligently in a 2^6 Factorial

| | | | A_1 | | | | A_2 | | | |
| | | | B_1 | | B_2 | | B_1 | | B_2 | |
			C_1	C_2	C_1	C_2	C_1	C_2	C_1	C_2
D_1	E_1	F_1	1	3	3	1				
		F_2	3	2	2	3	1			1
	E_2	F_1		1	1					
		F_2	2			2		1	1	
D_2	E_1	F_1					1			1
		F_2	1		1					
	E_2	F_1						1	1	
		F_2		1	1					

Table 7.34 Two Latin Squares with 3 Factors, Each at 2 Levels

	A_1	A_2
B_1	C_1	C_2
B_2	C_2	C_1

and

	A_1	A_2
B_1	C_2	C_1
B_2	C_1	C_2

Trials marked 2 are again a small Latin square with A_1, D_1, and F_2 constant. Trials marked 3 are a Latin square with A_1, D_1, and E_1 constant. Further averages can now be run for effects of B, C, E, and F.

OFFSETTING ERRORS FROM UNCONTROLLED OR UNCONTROLLABLE VARIABLES. This portion of the argument applies to both incomplete blocks or complete blocks run in selected segments. It so happens that for a factorial experiment of 3 factors, each at 2 levels, we have two possible Latin squares represented (see Table 7.34). Now, assuming that we feel *we must run all of the factorial plan*, but can run only four experiments at a time, what experiments must we choose to run together?

The answer depends upon where we want the error to lie. Or, in other words, where do we wish the error to affect us, assuming one set of data is in error or even assuming both sets of data are in error.

For a moment, let us choose to run each of the four experiments for the Latin squares together. Let us assume that the values obtained for the first Latin square in Table 7.34 are all high by six units and that the values obtained for the second Latin square are all low by eight units. Now we tabulate our full factorial (Table 7.35) *and enter only the errors*, $+6$ or -8 as the case may be. Now let us determine the errors for evaluation of A_1 versus A_2. This is the sum of the A_1 columns versus the A_2 columns.

$$+6 -8 -8 +6 \text{ versus } -8 +6 +6 -8$$
$$-4 \text{ versus } -4$$

The errors have canceled out.

Table 7.35 Factorial Based on Table 7.34, Entering the Errors Only

	A_1		A_2	
	C_1	C_2	C_1	C_2
B_1	$+6$	-8	-8	$+6$
B_2	-8	$+6$	$+6$	-8

The errors for evaluation of B_1 versus B_2 are the sums of the appropriate lines, or:

$$+6 -8 -8 +6 \text{ versus } -8 +6 +6 -8$$
$$-4 \text{ versus } -4$$

Again, the errors have canceled out. The same effect is true for evaluation of C_1 versus C_2.

This shows that the errors cancel out for each evaluation. *This is a case in point where perhaps different raw materials (a fourth variable) were used for each of the Latin squares, but such differences did not affect our final conclusions.*

What does confuse here is the effect when all variables at their low level $(A_1 \ B_1 \ C_1 = +6)$ are compared with all variables at their high level $(A_2 \ B_2 \ C_2 = -8)$. The comparison cannot be made (with the above data) independent of the error. In this case *another choice of four to be run together* is necessary. Take, for instance, those in Table 7.36. Both are required to evaluate C, but only the first may answer our question above. This is a step toward completing the factorial plan.

Table 7.36   Evaluating C

		A_1	A_2
	B_1	C_1	C_2
	B_2	C_1	C_2
and			
	B_1	C_2	C_1
	B_2	C_2	C_1

Now take a case where we are really not much interested in the effect of B_1 versus B_2. Then, we choose to hold B constant for each of our sets of four. Perhaps the data (errors) are as expressed in Table 7.37, but this is a pile-up for comparison of B_1 versus B_2.

Table 7.37   Comparison of B_1 and B_2 Errors

	A_1		A_2	
	C_1	C_2	C_1	C_2
B_1	-6	-6	-6	-6
B_2	$+8$	$+8$	$+8$	$+8$

This same type of selection of small segments of large plans to be run together can generally be done. Take, for instance, the plan where each of five factors, A, B, C, D, E, is to be studied at two levels. This is a factorial experiment requiring 32 tests. Suppose our raw material came in unit lots with just enough in each lot for 8 experiments. Now we have to use four different lots. The tabular layout is shown in Table 7.38. Perform together, *with same raw material:*

1st those marked ×	raw material lot (1)
2nd those marked ○	raw material lot (2)
3rd those marked +	raw material lot (3)
4th those marked *	raw material lot (4)

Table 7.38 Blocking of Segments of Large (2^5) Factorial Experiment for Use of Four Lots of Raw Material

		A_1				A_2			
		B_1		B_2		B_1		B_2	
		C_1	C_2	C_1	C_2	C_1	C_2	C_1	C_2
D_1	E_1	×	*	*	×	+	○	○	+
	E_2	○	+	+	○	*	+	×	*
D_2	E_1	○	+	+	○	*	×	×	*
	E_2	×	*	*	×	+	○	○	+

Now, when you want to average any group for effects of a certain factor, note that each average for the comparison contains equal numbers of each of the groups, ×, ○, +, and *.

Suppose we want to determine the effect of B_1 versus B_2. We take averages of data column 1, 2 versus 3, 4, and averages of data columns 5, 6, versus 7, 8. Note that *each of the averages* contains 2 values of ○, 2 values of ×, 2 values of * and 2 values of +. Now, when we make our comparisons, we have a balancing of errors for any grouping (○, ×, *, or +); the raw material lot is offset.

Consistent errors in any one group (○, ×, *, or +) *are ineffective* in overall averages, *no matter for which variable you take them.*

7.7 OTHER INCOMPLETE PLANS ILLUSTRATED

We have illustrated the plan which is $\frac{1}{2}$ of a 2^3 factorial in Table 7.20 (four experiments out of eight). One-half of a 2^4 factorial plan is illustrated in Table 7.39.

Table 7.39 One half of a 2^4 Plan

		A_1		A_2	
		B_1	B_2	B_1	B_2
C_1	D_1	*			*
	D_2		*	*	
C_2	D_1		*	*	
	D_2	*			*

We illustrate a plan for eight observations per block in a 2^5 factorial plan. The numbers indicate the block experiments performed together. Note that experiments marked 1 are smaller Latin squares, as are the others 2, 3, 4 (Table 7.40).

Table 7.40 Eight Observations per Block in a 2^5 Factorial Plan. Numbers Indicate Experiments Performed Together

		A_1				A_2			
		B_1		B_2		B_1		B_2	
		C_1	C_2	C_1	C_2	C_1	C_2	C_1	C_2
D_1	E_1	1	3	3	1	2	4	4	2
	E_2	4	2	2	4	3	1	1	3
D_2	E_1	2	4	4	2	1	3	3	1
	E_2	3	1	1	3	4	2	2	4

Table 7.41 Incomplete Block Plan of 12 Experiments (Ref. 26, Plan 1, p. 13–8)

Replications	Block	Treatments			
		A	B	C	D
I	1	×	×		
	2			×	×
II	3	×		×	
	4		×		×
III	5	×			×
	6		×	×	

Table 7.42 Incomplete Block Plan of 20 Experiments (Ref. 26, Plan 2, p. 13–8)

Replications	Block	Treatments				
		A	B	C	D	E
Four A's;	1	×	×			
Four B's; etc.	2		×			×
	3			×	×	
	4	×			×	
	5			×		×
	6	×		×		
	7		×		×	
	8		×	×		
	9				×	×
	10	×				×

Table 7.43 Incomplete Block Plan of 30 Experiments (Ref. 26, Plan 3, p. 13–9)

Replications	Block	Treatments					
		A	B	C	D	E	F
I	1	×	×				
	2			×	×		
	3					×	×
II	4	×		×			
	5		×			×	
	6				×		×
III	7	×			×		
	8		×				×
	9			×		×	
IV	10	×				×	
	11		×		×		
	12			×			×
V	13	×					×
	14		×	×			
	15				×	×	

Note in Table 7.40 that each data column contains data from blocks 1, 2, 3, 4. Note that each data line contains two results each for 1, 2, 3, 4. Note that the two diagonals for each half contain results for 1, 2, 3, 4. Striking averages here is obvious.

We give three other incomplete block plans (Tables 7.41, 7.42, 7.43) involving considerable replication. Usually we must select some repeat experiments and this can be done intelligently. Here the size of the block is not large enough to accommodate all treatments in one block. Further plans are listed in Reference 26; in fact nineteen are so tabulated up to a total of ten treatments, forty-five blocks, and nine replications.

8

ANALYSIS OF EXPERIMENTAL DATA

The average family in the United States has
2.4 children! Is your family average?

8.1 WHAT DO THE RESULTS MEAN?

You have just set up new equipment. You have run one result on an unknown material. What does it mean?

- Probably nothing.

You have just set up new equipment. You have run one result on an unknown material. You are sure the result is reliable. What does it mean?

- You are prejudiced. Result is wishful; it may be okay, but it probably means nothing. If correct, you were just lucky.

You have just set up new equipment. You run one result on a standard or known material and it checks the known value. What does it mean?

- It is an indication that the equipment is probably okay. But why not run a series on this known material?

You run the series desired above. The results cluster about the accepted value and show a normal distribution. What does it mean?

- Now you have proof that the equipment was okay on the day your run was made. Why not repeat it again next week if you have reason to doubt it?

You run the second week tests as described above. The results check the first ones within a reasonable amount. (This is to be described later.) What does it mean?

- Now you can trust your equipment, your procedure, and yourself.

You have proven equipment and procedure. You run an unknown material and get one result. What does it mean?

- Result is probably accurate (that is, 2 out of 3 times) to within one sigma (σ, to be explained) of the average of a large number of such tests.

You have proven equipment and procedure. You run an unknown material with a known material, whose result was as expected. What does it mean?

- Result on unknown material will be trustworthy, but not necessarily any closer than one sigma to the average (2 out of 3 times).

You proceed with a large number of runs. What does result mean?

- Your average is within (2 out of 3 times) sigma/$n^{\frac{1}{2}}$ of the true average of an infinitely large number of runs, where n = number of runs. But you also can estimate sigma now.

8.2   SIGMA

Sigma is purely a measure of the scatter of your results from the average value based on statistical concepts.

Sigma is that value, plus and minus, from the average of the population, which includes 68% of the results you obtain, or of the results which would be obtained if your sample were infinitely large. In other words, one result has 68 chances out of 100 of falling within 1 sigma from the average.

Sigma is a measure of spread, or of precision of measurement.

Sigma is the standard deviation of a series of data.

The sigma for a total population is abbreviated σ' (sigma prime).

The sigma for a sample population is abbreviated σ (sigma).

8.3   RESULTS WITH NORMAL DISTRIBUTION

You have all seen a normal distribution curve such as is shown in Figure 8.1. This is an ideal curve based on 10,000 pieces and shows the split of parts as a function of sigma (σ'), where the average $\bar{X}'$ is 20 and σ' is 2.816.

Note that from $+1\sigma'$ to $-1\sigma'$ (17.184 to 22.816) we have 6,827 pieces or (basis unity)

$+1\sigma'$ to $-1\sigma' = 0.6827$
$+2\sigma'$ to $-2\sigma' = 0.9545$
$+3\sigma'$ to $-3\sigma' = 0.9973$

or probability is:

0.683 that values lie within $\pm 1\sigma'$
0.954 that values lie within $\pm 2\sigma'$
0.997 that values lie within $\pm 3\sigma'$

That is, 683 out of 1,000 measurements lie within $\pm 1\sigma'$.

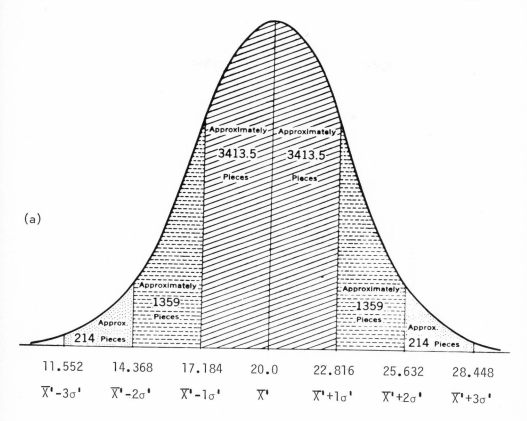

(a)

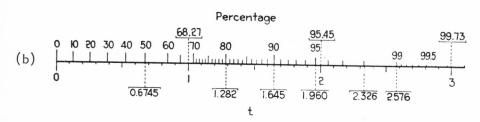

(b)

Figure 8.1 (a) The normal frequency distribution for 10,000 pieces. (b) Percentage of total area under a normal distribution curve, ± from the center of the population mean. Here t is used as a multiplier for σ in the equation Range $\bar{X}' = \bar{X}' \pm (t)(\sigma')$.

This perfect distribution can be represented as a straight line when plotted on "probability paper" as is done in Figure 8.2. Now we can easily read the cumulative percentages. For instance,

1% were under 13.3; 2% were under 14.1;
16% were under 17.18 (-1σ); 50% were under 20;
84% were under 22.82; 99% were under 26.7;
68.3% were between 17.18 and 22.82 ($\pm 1\sigma$)

The probability paper is so spaced as to make a linear plot when the distribution is normal. What a convenience! It follows conversely that if you plot data on this paper, then you can read off sigma without calculations.

8.4   EFFECT OF SAMPLE SIZE

We know that a small sample does not give the same results as a large sample. Now that we have a smattering of an idea of sigma, we wish to point out the effect of sample size on an expected average.

EXAMPLE E8-1.   We take a large universe which we know averages 200 and has a sigma of 20. Now we take smaller samples from 1 up to 100 and calculate the limits of the expected average. See Table 8.1. Here we have $\bar{X}'$ and σ'. The rule is that the σ for sample lots is calculated from σ', where n is the sample size, as follows:

$$\sigma = \frac{\sigma'}{n^{\frac{1}{2}}}$$

Table 8.1   Effect of Sample Size on Expected Average when Universe Average = $\bar{X}'$ = 200 and Sigma Prime = σ' = 20

Sample Size (n)	$\sigma'/n^{\frac{1}{2}}$	$\pm 1\sigma'/n^{\frac{1}{2}}$ (2 out of 3 times)	$\pm 3\sigma'/n^{\frac{1}{2}}$ (997 out of 1000 times)
		Limits of Expected Averages	
1	20	180–220	140–260
2	14	186–214	158–242
4	10	190–210	170–230
16	5	195–205	185–215
25	4	196–204	188–212
64	$2\frac{1}{2}$	$197\frac{1}{2}$–$202\frac{1}{2}$	$192\frac{1}{2}$–$197\frac{1}{2}$
100	2	198–202	194–206
Infinity	0	200–200	200–200

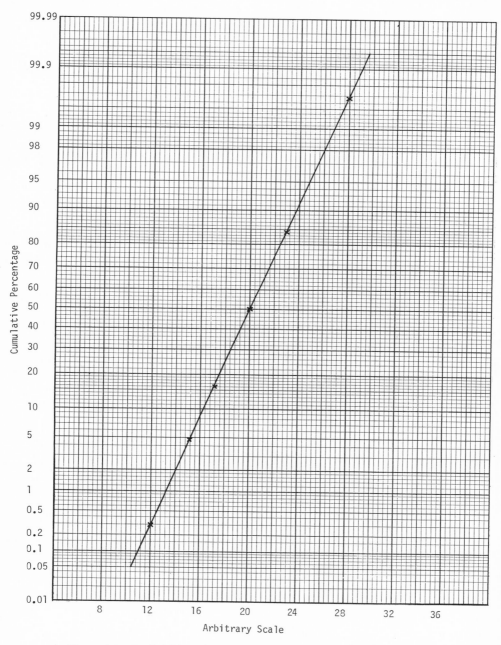

Figure 8.2 Probability paper plot of data from Figure 8.1.

This means that if we have a sample of 16, then 2 out of 3 times the average obtained will fall within 195 to 205, or 997 times out of 1000 the average will fall within 185 to 215. This is a very useful concept.

8.5   GRAPHICAL PROCEDURES USING PROBABILITY PAPER

Most engineers like to use graph paper, particularly if their results come out simply as a straight line. As noted in Section 8.3, the use of probability paper has several advantages over mathematical tabulation. However, care must be exercised in the procedure.

The methods used for this procedure are given in Table 8.2. We recommend that these procedures be attempted for all laboratory data.

How to Use Table 8.2

CASE 1.   Where approximately 30 or fewer measured values are involved, the following steps are taken:

1.   Arrange the data in order of increasing magnitude.
2.   Give each observation a cumulative percentage value from Table 8.2.
3.   Choose a proper scale, and plot each different value as the abscissa, and the cumulative percent as the corresponding ordinate.

EXAMPLE CASE 1 (illustrated).   Take the 25 observations listed here. Tabulate as instructed. Note that the lowest value is not given a cumulative percent of 4% although one observation is obviously 4% of the total number of observations (25). Each succeeding percentage is found by adding 4% to the preceding value.

Note how the three observations at 3.8 are plotted at 18, the midpoint value. It does not hurt to plot all three points.

The probability curve (opposite) approximates a straight line — the data is normally distributed. The average value, $\bar{X}$, of 4.8 is found at the intersection with the 50% line. The standard deviation, σ, is found to be 1.1 by subtracting the abscissa of the curve at 50% from the abscissa at 84%, or $5.9 - 4.8 = 1.1$.

CASE 2 (illustrated).   When a large number of observations is available, the process of plotting individual values is too lengthy. Thus, the following procedure is used:

1.   Tabulate the number of observations that lie in chosen equal-sized intervals. The interval should be at least 10 times the last significant figure to which the values are measured.
2.   Calculate the percentage of observations in each interval and the cumulative percentage in and below the interval (see tabulation).
3.   Plot the cumulative percent on the probability scale against *the top of the interval* on the abscissa.
4.   Interpret for average $\bar{X}$ and σ, as in Case 1, using the 50% and 84% values.

Table 8.2 Procedure for Graphic Plotting on Probability Paper (Ref. 21)

Obs. No.	Available Number of Observations																				
	10	11	12	13	14	15	16	17	18	19	20	21	22	23	24	25	26	27	28	29	30
1	5	4.5	4.2	3.8	3.6	3.3	3.1	2.9	2.8	2.6	2.5	2.4	2.3	2.2	2.1	2.0	1.9	1.9	1.8	1.7	1.7
2	15	13.6	12.5	11.5	10.7	10.0	9.4	8.8	8.3	7.9	7.5	7.1	6.8	6.5	6.25	6.0	5.8	5.6	5.4	5.2	5.0
3	25	22.7	20.8	19.2	17.8	16.7	15.6	14.7	13.9	13.2	12.5	11.9	11.4	10.9	10.4	10.0	9.6	9.3	8.9	8.6	8.3
4	35	31.8	29.2	26.9	25.0	23.3	21.9	20.6	19.4	18.4	17.5	16.7	15.9	15.2	14.6	14.0	13.5	13.0	12.5	12.1	11.7
5	45	40.9	37.5	34.6	32.1	30.0	28.1	26.4	25.0	23.7	22.5	21.4	20.4	19.6	18.75	18.0	17.3	16.7	16.1	15.5	15.0
6	55	50.0	45.8	42.3	39.2	36.7	34.4	32.3	30.6	29.0	27.5	26.2	25.0	23.9	22.9	22.0	21.2	20.4	19.6	19.0	18.3
7	65	59.1	54.2	50.0	46.4	43.3	40.6	38.2	36.1	34.2	32.5	30.9	29.6	28.3	27.1	26.0	25.0	24.1	23.2	22.4	21.7
8	75	68.2	62.5	57.7	53.5	50.0	46.9	44.1	41.7	39.5	37.5	35.7	34.1	32.6	31.25	30.0	28.9	27.8	26.8	25.9	25.0
9	85	77.3	70.8	65.4	60.7	56.7	53.1	50.0	47.2	44.8	42.5	40.5	38.7	37.0	35.4	34.0	32.7	31.5	30.4	29.3	28.3
10	95	86.4	79.2	73.1	67.8	63.3	59.4	55.9	52.8	50.0	47.5	45.2	43.2	41.3	39.6	38.0	36.6	35.2	33.9	32.8	31.7
11		95.5	87.5	80.8	75.0	70.0	65.6	61.8	58.4	55.3	52.5	50.0	47.7	45.7	43.75	42.0	40.4	38.9	37.5	36.2	35.0
12			95.8	88.5	82.1	76.7	71.9	67.7	63.9	60.6	57.5	54.7	52.2	50.0	47.9	46.0	44.2	42.6	41.1	39.7	38.3
13				96.2	89.2	83.3	78.1	73.6	69.4	65.8	62.5	59.5	56.8	54.4	52.1	50.0	48.1	46.3	44.6	43.1	41.7
14					96.4	90.0	84.4	79.5	75.0	71.1	67.5	64.4	61.4	58.7	56.25	54.0	51.9	50.0	48.2	46.6	45.0
15						96.7	90.6	85.4	80.6	76.4	72.5	69.1	66.0	63.1	60.4	58.0	55.8	53.7	51.8	50.0	48.3
16							96.9	91.2	86.1	81.6	77.5	73.8	70.5	67.4	64.6	62.0	59.6	57.4	55.4	53.5	51.7
17								97.1	91.7	86.9	82.5	78.6	75.0	71.8	68.75	66.0	63.5	61.1	58.9	56.9	55.0
18									97.2	92.2	87.5	83.4	79.6	76.1	72.9	70.0	67.3	64.8	62.5	60.4	58.3
19										97.4	92.5	88.1	84.1	80.5	77.1	74.0	71.2	68.5	66.1	63.8	61.7
20											97.5	93.0	88.6	84.8	81.25	78.0	75.0	72.2	69.6	67.3	65.0
21												97.7	93.2	89.2	85.4	82.0	78.8	75.9	73.2	70.7	68.3
22													97.7	93.5	89.6	86.0	82.7	79.6	76.8	74.2	71.7
23														97.9	93.75	90.0	86.5	83.3	80.4	77.6	75.0
24															97.9	94.0	90.4	87.0	83.9	81.1	78.3
25																98.0	94.2	90.7	87.5	84.5	81.7
26																	98.1	94.4	91.1	88.0	85.0
27																		98.1	94.6	91.4	88.3
28																			98.2	94.9	91.7
29																				98.3	95.0
30																					98.3

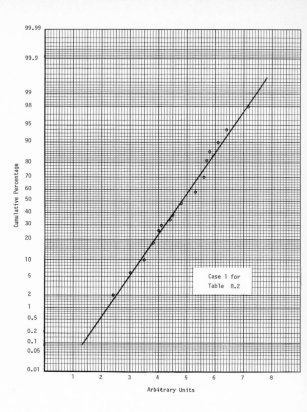

Case 1 for Table 8.2

Observed Value	Cumulative Percent
2.4	2
3.0	6
3.5	10
3.8	14
3.8	18
3.8	22
4.0	26
4.1	30
4.4	34
4.5	38
4.8	42
4.8	46
4.8	50
4.8	54
5.3	58
5.6	62
5.6	66
5.6	70
5.6	74
5.6	78
5.7	82
5.8	86
6.1	90
6.4	94
7.2	98

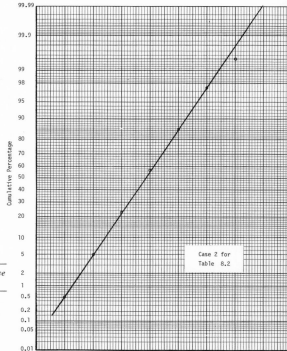

Case 2 for Table 8.2

Interval in Arbitrary Units	Number of Observations in Each Interval	Percent of Observations in Each Interval	Cumulative Percent
6.01–6.10	1	0.5	0.5
6.11–6.20	9	4.5	5.0
6.21–6.30	36	18.0	23.0
6.31–6.40	68	34.0	57.0
6.41–6.50	57	28.5	85.5
6.51–6.60	24	12.0	97.5
6.61–6.70	4	2.0	99.5
6.71–6.80	1	0.5	100.0

After plotting, you can *read off sigma directly*. Simply read off the value for 84% cumulative and subtract the value for 50% from it. The slope gives the sigma.

EXAMPLE E8-2.   Figure 8.3 gives a plot of three experiments on the manufacture of glass tubing. Variable A gave a sigma of 0.007. Read off 84% as 0.625 diameter and 50% as 0.618; the difference is 0.007, or sigma. Variable B gave a sigma of 0.005; Variable C gave 0.001. One should not pay attention to the average, as no effort was made to hold it constant.

EXAMPLE E8-3.   Figure 8.4 gives a probability plot for determination of sulfide sulfur in one amber glass (12 duplicate determinations). Sigma is read as $0.0281 - 0.02759 = 0.00051$. Note here that 12 determinations gave only six points on the plot.

EXAMPLE E8-4.   Figure 8.5 shows the hydrostatic pressure strength of 1,000 bottles as an illustration of marked deviation from a straight line. This illustrates *skew* where a few bottles of very high strength cause this deviation. See reference books on calculation of skew (Ref. 3) and its effect on analysis techniques.

EXAMPLE E8-5.   *Large Number of Observations*.   Later, in Table 8.5, we give the results of 270 measurements of strength of brick. Plotting 270 points on probability paper is time consuming. These data are already grouped by cells in Table 8.5, but we can make a probability plot as follows. The data are grouped by equal sized cells, together with observed frequency for each cell. Cumulative frequency is calculated, and converted to % cumulative frequency (see Table 8.6). We then plot % cumulative frequency versus the top of the cell interval. The example for these data is Figure 8.6.

EXAMPLE E8-6.   It is always safe to attempt these plots. They may be quite revealing. Take the case of Figure 8.7 where we have plotted the thickness of a material for both the morning and afternoon of an experiment. The A.M. data show poor fit to a straight line, while the P.M. data have deviations as expected (curving at the ends in the expected directions). Was this improvement due to an improvement in control of the experiment as the day progressed?

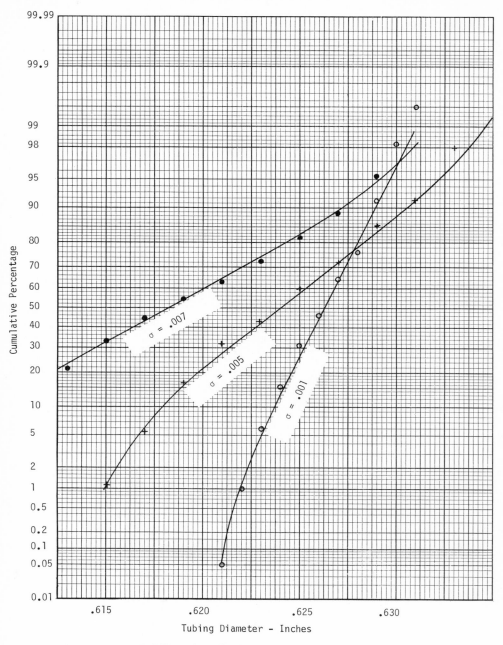

Figure 8.3 Probability plot of tubing diameter versus three variables, where ● stands for Variable A, + for Variable B, and ○ for Variable C. See Examples E8-2 and E10-1.

72

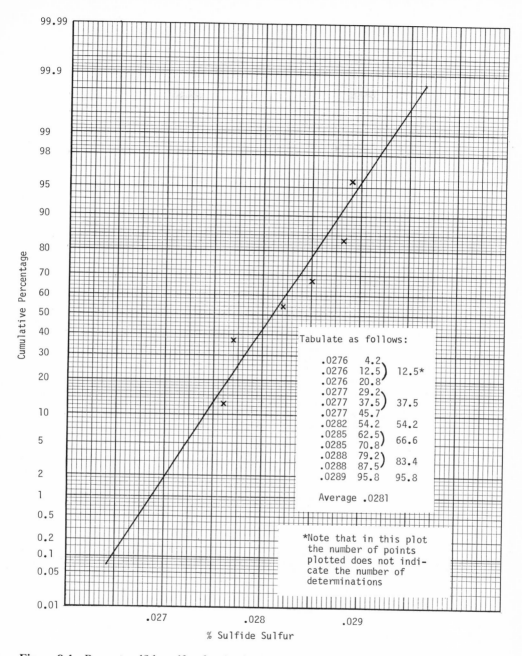

Figure 8.4 Percent sulfide sulfur for 12 determinations in one amber glass. See Examples E8-3 and E8-11. See Table 8.2 for procedure for plotting probability paper and tabulate as shown above. From Reference 9, P. Close, F. C. Raggon, and W. E. Smith, *Jour. Amer. Ceram. Soc.* 33:345, 1950, by permission of the American Ceramic Society. Copyright 1950 by the American Ceramic Society.

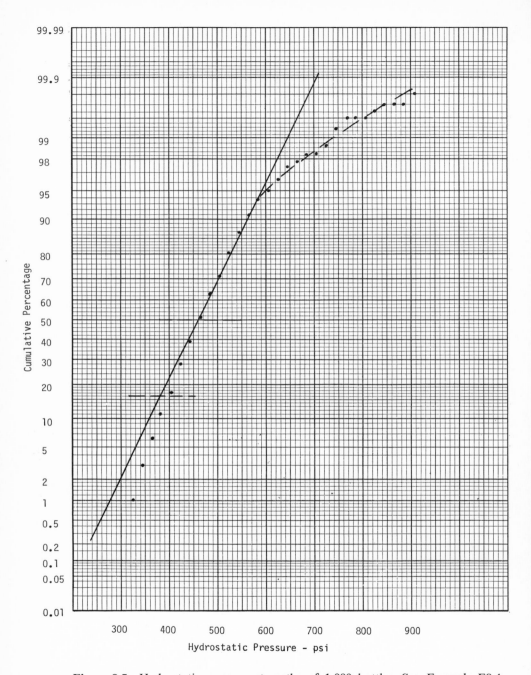

Figure 8.5 Hydrostatic pressure strengths of 1,000 bottles. See Example E8.4. This is an illustration of skew with calculated average 461 and sigma 74. From Reference 29, F. W. Preston, *Jour. Amer. Ceram. Soc.* 20:329, 1937, by permission of the American Ceramic Society. Copyright 1937 by the American Ceramic Society.

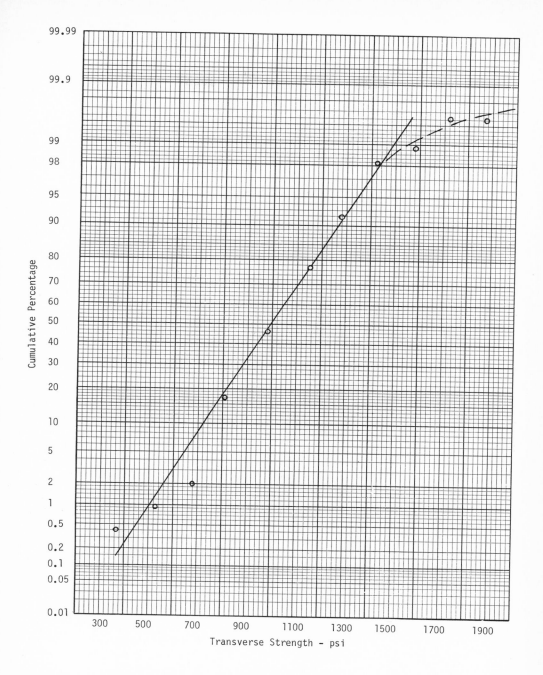

Figure 8.6 Strength of brick by cell plotting. See Examples E8-5 and E8-12.

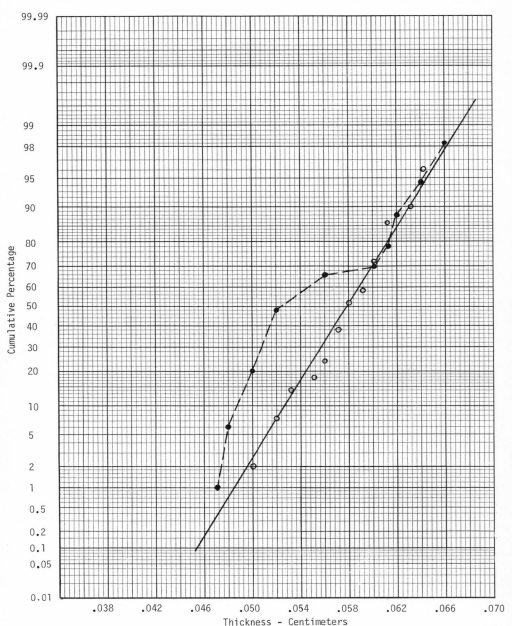

Figure 8.7 Thickness of a material run in A.M. versus P.M. of the same day; P.M. is indicated by ○, A.M. by ●. See Example E8-6.

8.6 MATHEMATICAL PROCEDURES FOR SIGMA

The standard deviation of a set of numbers is defined as the *root-mean-square (rms) deviation* of the numbers from their average:

$$\sigma = \left[\frac{(X_1 - \bar{X})^2 + (X_2 - \bar{X})^2 + \cdots + (X_n - \bar{X})^2}{n - 1}\right]^{\frac{1}{2}}$$

where $\bar{X}$ = arithmetic average; X_1, X_2, etc = individual data; n = number in set. If d = differences above determined, then

$$\sigma = \left[\frac{\sum (d^2)}{n - 1}\right]^{\frac{1}{2}} = \left(\frac{SS}{n - 1}\right)^{\frac{1}{2}}$$

where $\sum$ = summation and SS = sums of squares.

The procedure is to take deviations from average, square those, summate, divide by (number minus 1), and take the square root.

An *arithmetic shortcut* to avoid taking differences is to use the equation for summation of squares:

$$\sum (d^2) = X_1^2 + X_2^2 + \cdots + X_n^2 - \left[\frac{(X_1 + X_2 + \cdots + X_n)^2}{n}\right]$$

EXAMPLE E8-7.   Take the group

$$9, \quad 11, \quad 14, \quad 11, \quad 13, \quad 12, \quad 10, \quad 13, \quad 12, \quad 15$$

Total = 120   Average = 12   Sums of squares of differences = SS.

$$SS = (12 - 9)^2 + (12 - 11)^2 + \cdots + (12 - 15)^2 = 30$$

$$\sigma = \left(\frac{30}{9}\right)^{\frac{1}{2}} = (3.33)^{\frac{1}{2}} = 1.82$$

But also:

$$SS = (9)^2 + (11)^2 + \cdots + (15)^2 - \left[\frac{(9 + 11 + \cdots + 15)^2}{10}\right]$$

$$= 30 \quad \text{or} \quad \sigma = \left(\frac{30}{9}\right)^{\frac{1}{2}} = 1.82$$

These data yield easily to the probability chart plot (Figure 8.8).

A *second arithmetic shortcut* is as follows:

$$\sigma = \left(\frac{X_1^2 + X_2^2 + \cdots + X_n^2 - n\bar{X}^2}{n - 1}\right)^{\frac{1}{2}}$$

$$= \left(\frac{9^2 + 11^2 + \cdots + 15^2 - 12^2}{9}\right)^{\frac{1}{2}}$$

$$= \left(\frac{1470 - 1440}{9}\right)^{\frac{1}{2}} = (3.33)^{\frac{1}{2}} = 1.82$$

Shifting the zero point (scaling) is useful since σ depends upon differences.

Analysis of Experimental Data

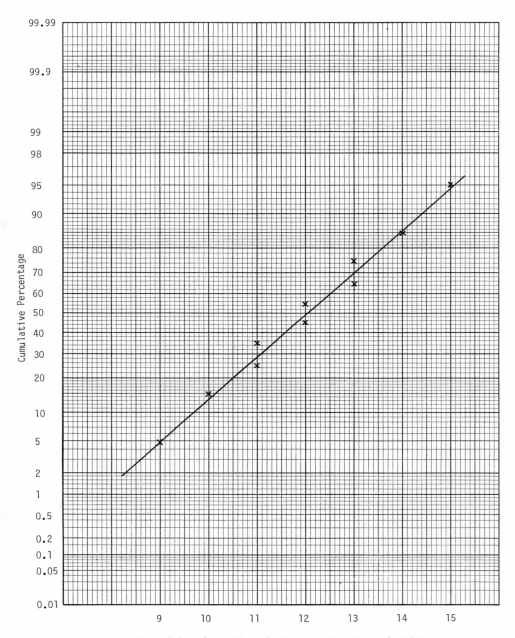

Figure 8.8 Plot of data for arithmetic shortcut. See Example E8-7.

EXAMPLE E8-8.   The group 109, 111, 114, 111, 113, 112, 110, 113, 112, 115 with total 1120 and average 112 has the same SS as in Example E8-7 above.

In a similar fashion, one may determine sigma on the numerator of a fraction only. Take the following example for weight of consecutively made products.

EXAMPLE E8-9, *Calculation of Sigma.*   Sixteen successive products weigh as shown in Table 8.3 (weight given only as numerator in $x/32$ pounds). From these data we get

$$\sigma = \left[\frac{\sum(\bar{x} - x)^2}{n - 1}\right]^{\frac{1}{2}} = \left(\frac{13.210}{15}\right)^{\frac{1}{2}} = (0.88)^{\frac{1}{2}} = 0.94$$

Also we have

$$\frac{30.0}{16} = 1.88$$

Adding 29, gives $\bar{X} = 30.88$. Average $= 30.88/32$ and $\sigma = 0.94/32$. But why do the above? These data yield easily to the probability plot (Figure 8.9).

Table 8.3   Calculating Sigma for 16 Successive Product Weights

Weight X	Subtract 29	$\bar{X} - X$	$(\bar{X} - X)^2$
30.5	1.5	0.38	0.144
31	2.0	0.12	0.014
32.5	3.5	1.62	2.624
32.5	3.5	1.62	2.624
31.5	2.5	0.62	0.384
30.0	1.0	0.88	0.774
30.5	1.5	0.38	0.144
31.5	2.5	0.62	0.384
32	3.0	1.12	1.254
31	2.0	0.12	0.014
29	0	0	0
30	1.0	0.88	0.774
32	3.0	1.12	1.254
30.5	1.5	0.38	0.144
30	1.0	0.88	0.774
29.5	0.5	1.38	1.904
Totals	30.0		13.210

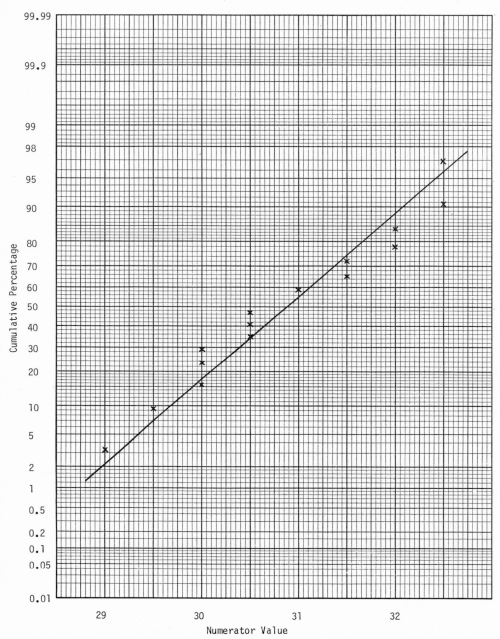

Figure 8.9 Using numerator only for probability plot. See Example E8-9.

Dividing or multiplying a set by a constant has the effect of dividing or multiplying the standard deviation by that constant.

EXAMPLE E8-10.   The group

90,   70,   70,   80,   20,   30,   70,   70,   90,   80

gives (Group 1) average = 67 and σ = 28.1.

These data are too scattered for a satisfactory probability chart plot. We have plotted the line (Figure 8.10) from our calculated average and calculated sigma.

The group 9, 7, 7, 8, 2, 3, 7, 7, 9, 8 gives average 6.7 and σ 2.81 (Group 2). See our double scaling in Figure 8.10.

EXAMPLE E8-11.   Take 12 analyses on sulfide sulfur in an amber glass (Table 8.4).

Table 8.4   Calculation of Sigma for Sulfide Sulfur in Amber Glass

Analysis	Subtract 0.0276 and Multiply by 10,000	d*	d^2
0.0276	0.	5.3	28.1
0.0276	0.	5.3	28.1
0.0276	0.	5.3	28.1
0.0277	1.	4.3	18.5
0.0277	1.	4.3	18.5
0.0277	1.	4.3	18.5
0.0282	6.	0.7	0.49
0.0285	9.	3.7	13.7
0.0285	9.	3.7	13.7
0.0288	12.	6.7	44.9
0.0288	12.	6.7	44.9
0.0289	13.	7.7	59.3
Totals	64.		316.8

*d is difference from column 2 and average of 5.3.

Total 64. $\sum d^2 = 316.8$ average 5.3

$$\sigma = \frac{1}{10,000} \left(\frac{316.8}{11}\right)^{\frac{1}{2}} = 0.00053$$

Average $= 0.0276 + 0.00053 = 0.0281$

But why do all this? See the probability plot Figure 8.4.

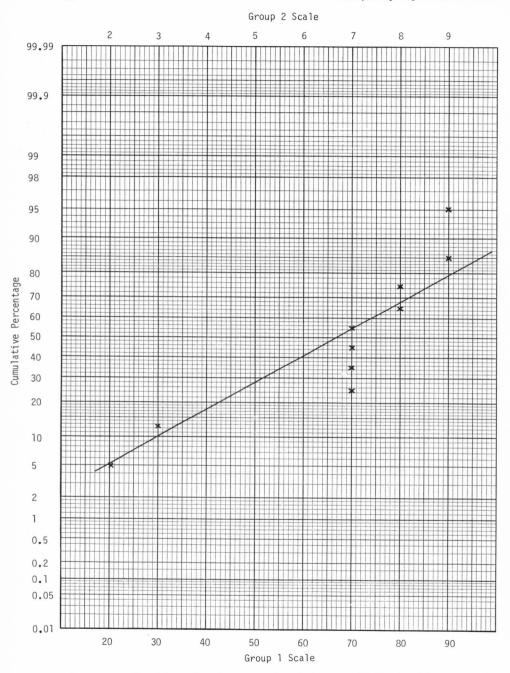

Figure 8.10 Double scaling (bottom and top) shows multiplication of average and sigma. See Example E8-10.

Calculating by cell number for a mass of data can be illustrated by the data in Table 8.5 (Ref. 3, p. 6).

Table 8.5   Strength of Brick by Cells

Group	Midpoint	Observed Frequency	Cumulative Frequency	% Cumulative Frequency
225–375	300	1	1	0.4
375–525	450	1	2	0.8
525–675	600	6	8	3.0
675–825	750	38	46	17.1
825–975	900	80	126	46.7
975–1125	1050	83	209	77.4
1125–1275	1200	39	248	91.9
1275–1425	1350	17	265	98.2
1425–1575	1500	2	267	98.9
1575–1725	1650	2	269	99.6
1725–1875	1800	0	269	99.6
1875–2025	1950	1	270	100.
		270		

EXAMPLE E8-12.   Assign letter A to the midpoint of the first cell. Let m equal the cell interval and f equal the observed cell frequency; x = deviation in cells from A. Calculate as in Table 8.6 where:

$$\bar{X} = A + m\left(\frac{\sum fx}{n}\right) \quad \text{and} \quad A = 300 \quad \text{and} \quad m = 150$$

$$\sigma = m\left[\frac{\sum fx^2}{n-1} - \frac{(\sum fx)^2}{n(n-1)}\right]^{\frac{1}{2}}$$

where

A = midpoint of cell No. 0 = 300 m = the cell interval = 150

Computations:

$$\bar{X} = A + m\left(\frac{\sum fx}{n}\right) = 300 + 150\left(\frac{1260}{270}\right)$$

$$= 300 + 150\,(4.67)$$

$$= 1000$$

$$\sigma = m\left[\frac{\sum fx^2}{269} - \frac{(\sum fx)^2}{(270)(269)}\right]^{\frac{1}{2}} = 150\left[\frac{6370}{269} - \frac{(1260)^2}{(270)(269)}\right]^{\frac{1}{2}}$$

$$= 150(23.70 - 21.86)^{\frac{1}{2}} = 202$$

This method gives a very close approximation.
But why do this? See the probability plot Figure 8.6.

Table 8.6   Calculation of Sigma by Cells (Strength of Brick)

Cell No.	Cell Midpoint	Deviation in Cells from A x	Observed Frequency f	fx	fx^2
0	A300	0	1	0	0
1	450	1	1	1	1
2	600	2	6	12	24
3	750	3	38	114	342
4	900	4	80	320	1280
5	1050	5	83	415	2075
6	1200	6	39	234	1404
7	1350	7	17	119	833
8	1500	8	2	16	128
9	1650	9	2	18	162
10	1800	10	0	0	0
11	1950	11	1	11	121
Total			n 270	Σfx 1260	Σfx^2 6370

8.7   SIGMA WITH INCOMPLETE DATA ON ANY SPECIFIC SAMPLE

We have indicated earlier that pairs of data, if available on a sufficiently large group of samples, are useful for calculation of sigma. This cannot be done graphically.

CALCULATION FROM PAIRS OF DATA ON DIFFERENT MATERIALS OR LOTS. The determination of two values is a common engineering concept of experimentation. Duplicate sets of any variety will allow calculation of sigma but not necessarily the average. It is reasonable to assume that variability of a test is constant even though the average value changes. Sometimes variability of data is proportional to the magnitude of the response; this is an exception for our calculation of σ.

The procedure is to take the differences (of duplicates), square, summate, divide by 2 times the number of pairs, and take the square root

$$\sigma = \left[\frac{\text{Sum of (differences)}^2}{2n}\right]^{\frac{1}{2}}$$

where n = number of *pairs*. This is equivalent to averaging each pair and proceeding normally as in the next section.

$$\sigma = \left\{\frac{\sum[(\bar{X} - X_1)^2 + (\bar{X} - X_2)^2 \text{ for each pair}]}{\text{number of values}}\right\}^{\frac{1}{2}}$$

Note here it is not necessary to strike an average for the pair. That is the reason that the factor 2 occurs in the denominator. Each pair of data really gives two differences to be squared, but when no average is struck, the factor 2 produces the same result. If W_1 and W_2 are a pair, then $\bar{W}$ is their average, and $(W_1 - \bar{W})^2 + (\bar{W} - W_2)^2$ is the same as $(W_1 - W_2)^2/2$.

EXAMPLE E8-13. We have a set of duplicate runs on Fe_2O_3 in sand. Calculate the standard deviation, σ, using Table 8.7.

Table 8.7 Sigma from Pairs of Data. Iron Oxide in Sand

Data		d	d^2
0.0043	0.0047	0.0004	0.000000
0.014	0.012	0.002	0.000004
0.033	0.034	0.001	0.000001
0.107	0.107	0.000	0.000000
0.036	0.035	0.001	0.000001
0.062	0.060	0.002	0.000004
0.099	0.103	0.004	0.000016
0.019	0.021	0.002	0.000004
0.079	0.080	0.001	0.000001
0.018	0.018	0.000	0.000000
0.036	0.036	0.000	0.000000
0.029	0.029	0.000	0.000000
0.035	0.033	0.002	0.000004
0.037	0.040	0.003	0.000009
0.026	0.026	0.000	0.000000
0.026	0.025	0.001	0.000001

Average difference = 0.0012 $\sum d^2 = 0.000045$

$$\sigma = \left[\frac{0.000045}{2(16)} \right]^{\frac{1}{2}} = (0.00000140)^{\frac{1}{2}} = 0.0012$$

Thus a single determination is likely to be within 0.0012 of the true value roughly 2 out of 3 times. Similarly, the average of a pair is likely to be within $0.0012/\sqrt{2} = 0.0009$ (see Table 8.1).

EXAMPLE E8-14. We have several sets of duplicate runs on Na_2O in glass. Calculate the sigma.

$(14.53 - 14.44)\ (14.80 - 14.69)\ (14.15 - 13.93)\ (15.05 - 14.94)$
$(\ 7.54 -\ \ 7.51)\ (17.68 - 17.64)\ (18.43 - 18.36)\ (15.41 - 15.29)$
$(15.26 - 15.38)\ (18.88 - 18.87)\ (13.04 - 13.00)\ (14.65 - 14.58)$
$(14.67 - 14.76)\ (12.34 - 12.49)\ (13.57 - 13.60)\ (12.86 - 12.92)$

Take the differences and square and summate:

$0.09^2 + 0.11^2 + 0.22^2 + 0.11^2 + 0.03^2 + 0.04^2 + 0.07^2 + 0.12^2 + 0.12^2$
$+ 0.01^2 + 0.04^2 + 0.07^2 + 0.09^2 + 0.15^2 + 0.03^2 + 0.06^2 = 0.1586 = SS$
of differences

$$\sigma = \left[\frac{0.1586}{2(16)}\right]^{\frac{1}{2}} = (0.00496)^{\frac{1}{2}} = 0.0704$$

$$3\sigma = 0.211$$

Essentially all of the values are within 0.21 of actual content of Na_2O. Of actual content of Na_2O, 2/3 of values are within 0.07. Average of 1 pair is within $0.07/\sqrt{2} = 0.05$ of actual content.

CALCULATION FOR SEVERAL INDEPENDENT LOTS OF DATA. If the σ values of several different sets are known, then we may calculate an average σ for all the sets, as follows:

$$\sigma \text{ all sets} = \left[\frac{\sigma_1^2(n-1) + \sigma_2^2(n_2-1) + \cdots + \sigma_k^2(n_k-1)}{n_1 + n_2 + \cdots + n_k}\right]^{\frac{1}{2}}$$

Actually, by squaring each σ and multiplying it by the number of measurements involved, we are rebuilding the sums of squares which we originally calculated. These SS are then summated, divided by the sums of the number of measurements, and the square root taken.

EXAMPLE E8-15. We have four groups of five determinations each on softening point of glass (Ref. 24). Calculate the standard deviation (Table 8.8).

$$\sigma = \left(\frac{0.71 + 0.98 + 1.52 + 1.48}{20}\right)^{\frac{1}{2}} = (0.2345)^{\frac{1}{2}} = 0.484$$

$$3\sigma = 1.45°C$$

Thus we have great confidence that the value is within 1.45°C of the true value.

Note: This same type of calculation applies also for cases where the groups are of uneven size.

8.8 LIMITS OF UNCERTAINTY OF AN OBSERVED AVERAGE

Confidence — that cocky feeling you have
just before you know better.

Theoretically, if our small sample is representative of the population, then the calculation is simple. But such is not usually the case. As a rule the procedures are then somewhat empirical. The theoretical effects of sample

Table 8.8   Calculation of Sigma from Groups of Data. The Softening Point of Glass (from Ref. 24)*

		d	d^2	
Group 1	623.6	0.5	0.25	
	623.0	0.1	0.01	$\sigma_1 = 0.422$
	623.6	0.5	0.25	
	622.9	0.2	0.04	$\sigma_1^2(n-1) = 0.71$
	622.7	0.4	0.16	
Average	623.1		0.71	
Group 2	696.3	0.2	0.04	
	696.0	0.1	0.01	$\sigma_2 = 0.495$
	695.3	0.8	0.64	$\sigma_2^2(n-1) = 0.98$
	696.3	0.2	0.04	
	696.6	0.5	0.25	
Average	696.1		0.98	
Group 3	818.4	0.5	0.25	$\sigma_3 = 0.616$
	818.4	0.5	0.25	
	818.8	0.1	0.01	$\sigma_3^2(n-1) = 1.52$
	819.9	1.0	1.00	
	818.8	0.1	0.01	
Average	818.9		1.52	
Group 4	922.1	0.7	0.44	
	923.3	0.5	0.25	
	922.7	0.1	0.01	$\sigma_4 = 0.608$
	923.6	0.8	0.64	
	922.5	0.3	0.09	$\sigma_4^2(n-1) = 1.48$
Average	922.8		1.48	

size were illustrated in Table 8.1, where for a probability of 997 times out of 1,000 we used the formula:

$$\text{Limits} = \bar{X} - \frac{3\sigma'}{n^{\frac{1}{2}}} \text{ to } \bar{X} + \frac{3\sigma'}{n^{\frac{1}{2}}}$$

CONFIDENCE LIMITS USING K TABLE.   Due to the sampling process, the estimates of average and sigma vary from sample to sample. If we have a random sample from a normal population, we can make interval estimates

*From J. T. Littleton, *Journal of the Society of Glass Technology*, 24:180, 1940, by permission of the Society of Glass Technology. Copyright 1940 by the Society of Glass Technology.

of the average and sigma. The interval obtained may or may not bracket the true parameter values. We must choose a confidence level in calculating these intervals. Such intervals are known as confidence intervals, or confidence limits. Basically, we are aware that larger samples give narrower confidence intervals.

In Table 8.1, we illustrated the effect of sample size where $\bar{X}'$ and σ' were known. But we often do not know these values.

In control limit work — which we make no effort to discuss in this book — the usage of one sigma, two sigma, and three sigma terminology is commonly accepted. Thus our limits from an average $\bar{X}$ are frequently set as $\bar{X} + 3\sigma$, known as the UCL (upper control limit), and $\bar{X} - 3\sigma$, known as the LCL (lower control limit). Here the sample size is expressed in a set sampling procedure for that particular product and characteristic.

But we generally, in engineering work, will vary the sample size, n, and our equation for limits versus $\bar{X}$ and σ is broadened, as follows: high limit $= \bar{X} + t(\sigma/n^{\frac{1}{2}})$, low limit $= \bar{X} - t(\sigma/n^{\frac{1}{2}})$. It is convenient to say that $K = t/n^{\frac{1}{2}}$ and define t as a table value, common in statistical analysis, which varies with the confidence level, discussed above, and the sample size. Commonly the confidence level is chosen as 0.99 or 0.95, which means that the calculated interval would include the population parameter 99 times out of 100, or 95 times out of 100. The risk of rejection based on t is 1 or 5 out of 100, respectively.

The t tables are based, for our selection, on the 0.99 or 0.95, but there are two other important points to remember. First, the distribution curve has two tails, one at the lower end, and one at the upper end. Thus t tables generally are based on one end values only. That is, for a 5% risk of rejection, $2\frac{1}{2}\%$ is at one end, and $2\frac{1}{2}\%$ at the other.

Secondly, the t table values are stated versus degrees of freedom, df. We discuss this briefly later, Section 9.5. Suffice it to say here that for our present purpose, a sample of (n) size gives $(n - 1)$ degrees of freedom. The t table usage is more completely discussed later (see Section 9.4).

Table 8.9 gives the K values for three probability levels. Figure 8.11 plots these values for rough interpolation extended from 0.70 to 0.99 probability.

To get a *range* of the average which includes the population average $\bar{X}'$ we use the formula $\bar{X}' = \bar{X} \pm K\sigma$ for a sample lot of n observations, with average $\bar{X}$ and σ. We now go to some illustrations on the usage of K.

EXAMPLE E8-16. Assume 17 tests, average of test values 188, and $\sigma = 20$. For confidence of 95% $K = 0.514$. Population average is known as 200.

$$K\sigma = (0.514)(20) = 10.280$$

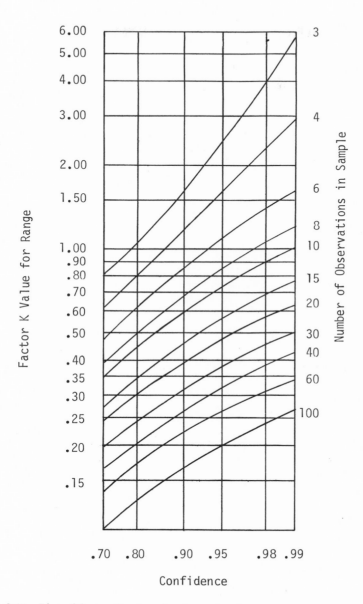

Figure 8.11 Plot of factor K values for determination of range for averages based on confidence of 0.70 to 0.99.

Range:

$$\bar{X} + 10.28 = 188 + 10 = 198$$
$$\bar{X} - 10.28 = 188 - 10 = 178$$

Range is to include $\bar{X}'$: 178 to 198.

Conclusion: The sample is not representative of the population, whose average ($\bar{X}'$) is 200.

EXAMPLE E8-17.   Assume 10 tests, average 575, with σ of 8. For confidence of 95% $K = 0.715$.

$$K\sigma = (0.715)(8) = 5.720$$

Range:

$$\bar{X} + 5.7 = 575 + 5.7 = 581$$
$$\bar{X} - 5.7 = 575 - 5.7 = 569$$

Range: 569 to 581

Conclusion: Here on a basis of 95% confidence, the $\bar{X}$ is 575, and the indicated range for $\bar{X}'$ is 581 to 569. Since we have no more data we cannot make a definite conclusion.

How to Use the *K* Table 8.9

Limits within which $\bar{X}'$ may be expected to lie (9 times in 10, 95 times in 100, or 99 times in 100) in a series of problems each involving a single sample of n observations. Confidence limits are $\bar{X} \pm K\sigma$.

To get a range for the average which includes the population average $\bar{X}'$

$$\bar{X}' = \bar{X} \pm K\sigma$$

The appropriate t value, double tail, from our Table 9.3, for 90% confidence = 0.10, for 95% confidence = 0.05, and for 99% confidence = 0.01.

For other n values, select t for $(n - 1)$ df and the appropriate confidence interval.

$$K = \frac{t_{(n-1)} \text{ for the appropriate confidence}}{n^{\frac{1}{2}}}$$

The 95% confidence column comes from Reference 32, Table 7 (K_1 values). The 90% and 99% confidence columns are calculated by the present author.

Table 8.9 Factors for K Values for Calculating Confidence Limits for Averages*

Sample Size	Values of K		
	90% Confidence	95% Confidence	99% Confidence
2	4.48	8.986	45.1
3	1.69	2.484	5.74
4	1.175	1.591	2.92
5	0.954	1.242	2.06
6	0.822	1.050	1.65
7	0.728	0.925	1.39
8	0.670	0.836	1.24
9	0.620	0.769	1.12
10	0.578	0.715	1.03
11	0.547	0.672	0.955
12	0.518	0.635	0.897
13	0.493	0.604	0.847
14	0.473	0.577	0.803
15	0.454	0.554	0.768
16	0.439	0.533	0.738
17	0.423	0.514	0.707
18	0.410	0.497	0.683
19	0.398	0.482	0.660
20	0.387	0.468	0.640
21	0.377	0.455	0.622
22	0.367	0.443	0.603
23	0.357	0.432	0.587
24	0.350	0.422	0.553
25	0.342	0.413	0.560
26	0.335	0.404	0.548
27	0.328	0.396	0.534
28	0.322	0.388	0.524
29	0.315	0.380	0.512
30	0.309	0.373	0.503
40	0.267	0.320	0.427
60	0.216	0.258	0.343
100	0.166	0.199	0.264
120	0.153	0.183	0.242
∞	0	0	0

*95% confidence column from Rochester Institute of Technology, *Symbols, Definitions and Tables for Industrial Statistics and Quality Control*, by permission of Eastman Kodak Company. Copyright 1958 by Eastman Kodak Company.

EXAMPLE E8-18.   In 25 tests upon the time interval in a measured process, we got the following (see full data in Table 9.5):

Average time	0.472 seconds
Sigma	0.006 seconds
Maximum time measured	0.483 seconds
Minimum time measured	0.463 seconds

What is the general range of average results if an extremely large sample were run?

$$\bar{X} = 0.472 \qquad n = 25 \qquad \sigma = 0.006$$

With 95% confidence $K = 0.413$; $K\sigma = (0.413)(0.006) = 0.0025$

$$\text{Range } \bar{X}' = 0.472 - 0.0025 = 0.469$$
$$= 0.472 + 0.0025 = 0.475$$

In other words, further data might have changed the average 0.0025 either way from the 0.472 value.

Note: In using $K\sigma$, remember that K is a multiplier and that $K\sigma$ is not a new value of σ itself.

THE GENERAL PROCEDURE FOR CONFIDENCE LIMITS.   The formula:

$$\text{confidence limits} = \bar{X} \pm \frac{t\sigma}{n^{\frac{1}{2}}}$$

The factor t is read for a very large sample (∞) from a two-tailed t table based on the desired probability. You can pick the probability you desire and calculate $K = t/n^{\frac{1}{2}}$. But we give K values directly in Table 8.9.

An interesting table (Table 8.10) for an infinitely large sample shows the range of t. You can see the range of K. The range of K thus tabulated is from 0.02 to 0.16, which shows that the measured range for $\bar{X}'$ is very narrowed for a sample of, say, 625 units. Suppose the $\bar{X}$ obtained on 625 units was 1000 and $\sigma = 50$. Then K and $\bar{X}'$ tabulate as shown in Table 8.11.

This table says that once out of twice, an average of 625 units will fall between 999 and 1001; if we want a higher confidence, then, say 99 times out of 100, that average with fall between 995 and 1005.

8.9   LIMITS OF UNCERTAINTY OF AN OBSERVED SIGMA

Our prime discussion of the uncertainty of an observed sigma is introduced in Section 10. But we cannot finish our discussion of the uncertainty of an observed average without some related discussion to be now presented.

While we do not need to go into the theory, tables of values for the lower limit and the upper limit of σ are available, based upon degrees of

Table 8.10   *t* Values for Uncertainty of Average, Large Sample Size

Confidence Level	t Value	If n = 625 K is	Rough Odds
0.5	0.674	0.025	1 out of 2
0.68	1.00	0.04	2 out of 3
0.75	1.15	0.046	3 out of 4
0.80	1.28	0.051	4 out of 5
0.90	1.64	0.066	9 out of 10
0.95	1.96	0.078	19 out of 20
0.96	2.05	0.082	96 out of 100
0.97	2.17	0.087	97 out of 100
0.98	2.33	0.093	98 out of 100
0.99	2.58	0.103	99 out of 100
0.9973	3.00	0.120	9973 out of 10,000
0.99983	4.00	0.160	99984 out of 100,000

Table 8.11   Population Range, $\bar{X}'$, when $\bar{X}$ on 625 units = 1000 and $\sigma = 50$

Confidence Level	K	Kσ	Population Range Indicated $\bar{X}'$
0.50	0.025	1.25	999–1001
0.90	0.066	3.30	997–1003
0.95	0.078	3.90	996–1004
0.99	0.103	5.15	995–1005
0.997	0.120	6.00	994–1006
0.9998	0.160	8.00	992–1008

freedom ($n - 1$). Our Table 8.12 gives the A_L (A_{Lower}) and the A_U (A_{Upper}) values which are used as multipliers for the σ measured from sample size n.

EXAMPLE E8-19.   In the Example E8-18, we had the data:

Average time $\bar{X}$	0.472 seconds
Sigma, σ	0.006 seconds
Sample size, n	25
Confidence selected	95%

Just how accurate is our sigma, or what are the predicted limits (upper and lower) which establish the predicted range of σ' from these data? For sample of 25 units, the table gives (interpolating roughly)

$$A_L = 0.78$$
$$A_U = 1.39$$

Table 8.12 Factors for A_L and A_U Values for Calculating Confidence Limits for Sigma. Adapted from Table A-21, Reference 26.

Sample Size	90% Confidence		95% Confidence		98% Confidence		99% Confidence	
	A_L	A_U	A_L	A_U	A_L	A_U	A_L	A_U
2	0.5103	15.947	0.4461	31.910	0.3882	79.786	0.3562	159.576
3	0.5778	4.415	0.5207	6.285	0.4660	9.975	0.4344	14.124
4	0.6196	2.920	0.5665	3.729	0.5142	5.111	0.4834	6.467
5	0.6493	2.372	0.5992	2.874	0.5489	3.669	0.5188	4.396
6	0.6721	2.089	0.6242	2.453	0.5757	3.003	0.5464	3.485
7	0.6903	1.915	0.6444	2.202	0.5974	2.623	0.5688	2.980
8	0.7054	1.797	0.6612	2.035	0.6155	2.377	0.5875	2.660
9	0.7183	1.711	0.6754	1.916	0.6310	2.204	0.6037	2.439
10	0.7293	1.645	0.6878	1.826	0.6445	2.076	0.6177	2.278
11	0.7391	1.593	0.6987	1.755	0.6564	1.977	0.6301	2.154
12	0.7477	1.551	0.7084	1.698	0.6670	1.898	0.6412	2.056
13	0.7554	1.515	0.7171	1.651	0.6765	1.833	0.6512	1.976
14	0.7624	1.485	0.7250	1.611	0.6852	1.779	0.6603	1.909
15	0.7688	1.460	0.7321	1.577	0.6931	1.733	0.6686	1.854
16	0.7747	1.437	0.7387	1.548	0.7004	1.694	0.6762	1.806
21	0.7979	1.358	0.7650	1.444	0.7297	1.556	0.7071	1.640
26	0.8149	1.308	0.7843	1.380	0.7511	1.473	0.7299	1.542
31	0.8279	1.274	0.7991	1.337	0.7678	1.416	0.7477	1.475
41	0.8470	1.228	0.8210	1.279	0.7925	1.343	0.7740	1.390
51	0.8606	1.199	0.8367	1.243	0.8103	1.297	0.7931	1.337
61	0.8710	1.179	0.8487	1.217	0.8239	1.265	0.8078	1.299
71	0.8793	1.163	0.8583	1.198	0.8349	1.241	0.8196	1.272
81	0.8861	1.151	0.8662	1.183	0.8449	1.222	0.8293	1.250
91	0.8919	1.141	0.8728	1.171	0.8515	1.207	0.8376	1.233
101	0.8968	1.133	0.8785	1.161	0.8581	1.195	0.8446	1.219

How to Use Table 8.12

Limits within which σ' may be expected to lie (9 times in 10, 95 times in 100, 98 times in 100, 99 times in 100) in a series of problems each involving a single sample of n observations. Confidence lower limit is $A_L\sigma$; Upper Limit is $A_U\sigma$.

From your sample size, read off A_L and A_U as multipliers for your measured sigma. Suppose σ measured is 14 for a sample size of 10 units. The limits for sigma are

Lower limit $= (\sigma)(A_L)$
Upper limit $= (\sigma)(A_U)$
Lower limit $= (14)(0.69)$ 0.95 confidence
Upper limit $= (14)(1.83)$ 0.95 confidence
Range $\sigma \quad = 9.7$ to 25.6

for 95% confidence. Note the table cuts off $2\frac{1}{2}$% at each end of top and bottom of distribution

$$\text{Lower } \sigma = 0.78\sigma = (0.78)(0.006) = 0.0047$$
$$\text{Upper } \sigma = 1.39\sigma = (1.39)(0.006) = 0.0083$$

Range of σ' is therefore predicted, in 95 times out of 100, as range σ' between 0.0047 and 0.0083.

Note particularly that the range predicted for σ' is different from the σ of the sample by a different amount for the upper and the lower values. Sigma cannot be negative, and skewness of sigma results from this prime factor. Thus

$$\text{Upper value } 0.0083 - 0.006 = 0.0023$$
$$\text{Lower value } 0.006 - 0.0047 = 0.0013$$

Note: You may now want to refer to Section 10 for a more complete discussion of testing the variability of sigma.

8.10 UNCERTAINTY VERSUS SAMPLE SIZE PLOTTED GRAPHICALLY

We have shown the uncertainty of the average and of sigma. Now, we wish to illustrate them by graphical plots on probability paper.

EXAMPLE E8-20. Assume a distribution calculated from 10 samples, using a 95% confidence on both the average and sigma.

$$\text{Measured } \bar{X} = 9.5 \qquad \text{Measured } \sigma = 2.0$$

K from Table 8.9 for 95% confidence is 0.715, so the predicted limits $\bar{X}'$ are $\pm K\sigma$ or

$$+(0.715)(2.0) = 1.4 \qquad \text{and} \qquad -(0.715)(2.0) = 1.4$$

or $\bar{X}'$ range is $9.5 \pm 1.4 = 8.1$ to 10.9. From Table 8.12, the A_U for 95% confidence is 1.826; A_L is 0.6878. Therefore

$$\text{Upper } \sigma = (1.826)(2.0) = 3.66 \qquad \text{Lower } \sigma = (0.6878)(2.0) = 1.38$$

Now we plot the range of $\bar{X}'$ at 50%, Figure 8.12; that is 8.1 to 10.9. To this we add at 1σ positions (84%) and (16%) the range of σ'; that is 1.38 to 3.66. These points are

at 84%	to the left	$9.5 + 1.38 = 10.9$
	to the right	$9.5 + 3.66 = 13.2$
at 16%	to the right	$9.5 - 1.38 = 8.1$
	to the left	$9.5 - 3.66 = 5.8$

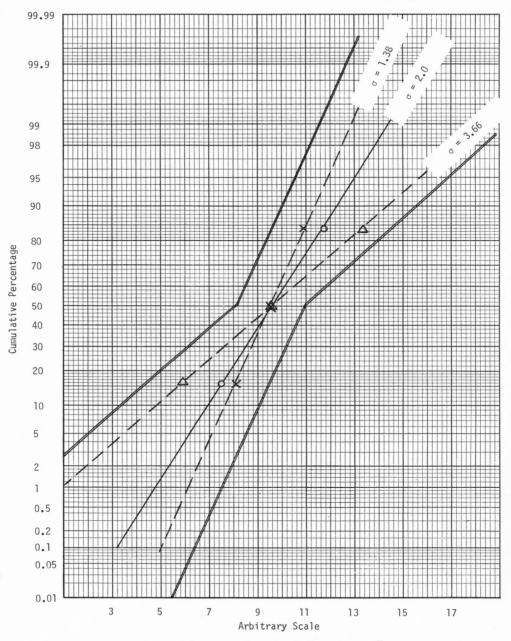

Figure 8.12 The confidence envelope (0.95) for $n = 10$, $\overline{X} = 9.5$, and $\sigma = 2.0$; calculated range of $\overline{X}' = 8.1$ to 10.9 and calculated range of $\sigma' = 1.38$ to 3.66. See Example E8-20.

In other words, as explained in Section 8.5, we draw lines for the population with the $\bar{X} = 9.5$ and the three σ values, 1.38, 2.0, 3.66. These lines are identified in Figure 8.12.

Now to get our entire possible range on the chart, we must take the limits as established for $\bar{X}'$ (that is 8.1 to 10.9) and draw similar curves. A convenient way to do this is to start at the extreme average limits and draw lines parallel to the nearest of the three original lines. Remember that parallel lines represent similar values of sigma.

The total inclusive area is that shown in Figure 8.12. This is a fan-shaped enclosure often identified as the *confidence envelope. Any linear line* drawn within this shaded area *could* represent the time $\bar{X}'$ and σ' situation. We see that the data of 10 samples are not very useful for any precise definition of our population.

EXAMPLE E8-21.   Let us assume the same $\bar{X}$ as above, 9.5, the same σ, 2.0, but also assume that these data resulted from a sample (n) of 100. We will now calculate as above indicated.

K from Table 8.9 for 95% confidence is 0.199.

Predicted $\bar{X}'$ limits are $\pm K\sigma$ or

$$+(0.199)(2.0) = 0.40 \qquad \text{and} \qquad -(0.199)(2.0) = 0.40$$

Thus $\bar{X}'$ range is $9.5 \pm 0.40 = 9.1$ to 9.9, and A_U for Table 8.12 for 95% confidence is 1.16; A_L is 0.879. Therefore

$$\text{Upper } \sigma = (1.16)(2.0) = 2.32$$
$$\text{Lower } \sigma = (0.879)(2.0) = 1.76$$

We plot these data in Figure 8.13. Note the great reduction in the area of the confidence envelope for a sample size of 100.

The increase in precision, by the way, is not linear with the number in the sample, but varies as the square root of the number of observations.

Certainty can be very expensive.

8.11   THE GRAPHICAL PLOT IN RELATION TO TOLERANCE OR SPECIFICATION

The average of a group and the standard deviation are a measure of quality of the process. It is highly important that the quality as produced has the proper relationship to the quality demanded. No process is commercial unless this is true. Specifications of final product must be based on what can be made. We do not make a product to have it later discarded. Furthermore, we must generally make it so that a very large share of the output reaches the required quality.

For instance, we may produce an item with an average strength of 180

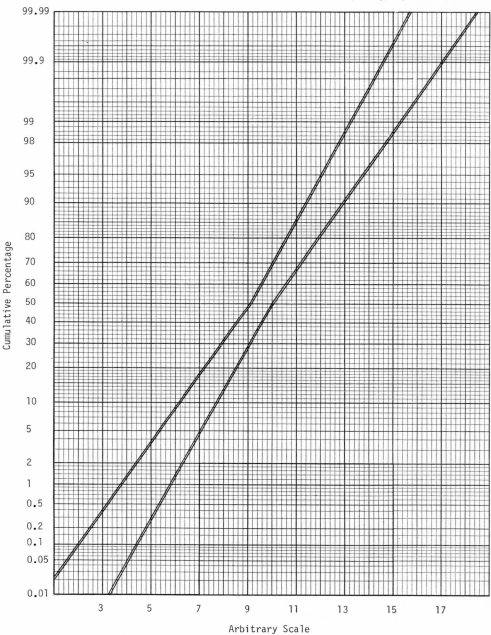

Figure 8.13   The confidence envelope (0.95) for $n = 100$, $\bar{X} = 9.5$, and $\sigma = 2.0$; calculated range of $\bar{X}' = 9.1$ to 9.9 and calculated range of $\sigma' = 1.76$ to 2.32. See Example E8-21.

units and a σ of 40 units. This means that 99.7% of the output falls within a range of 60 to 300 units; and that 68% of the output falls within 140 to 220 units. Now suppose that the consumer usage demanded a minimum of 220 units strength, either arbitrarily or factually. The data tell us that only about 1/6 of our product reaches this value. We have then either an un-economic process (because it does not produce a good enough product) or an arbitrary product standard which is uneconomic. The economics de-pends on manufacturing and selling costs. In fact there are cases where only half the production is salable, but a profit may still be made.

The predictions of the standard frequency curve in relation to variables measured are important. Figure 8.1 shows how 10,000 units break in rela-tion to the plus or minus σ' limits or values.

EXAMPLE E8-22.   Suppose that we can use an article only if a certain dimension were maintained between, say, 0.25 and 0.45. In producing to an average of 0.35 we have a sigma of 0.05.

Range $= 0.35 \pm 3\sigma = 0.20$ to 0.50

Available or useful range 0.25 to 0.45 or $+0.10$ and -0.10 from 0.35

Dividing ± 0.10 by $0.05 = \pm 2\sigma$ limits only

Obviously only 95% of the production can be used. The problem is definite — we must produce with more precision.

We next lower our σ by better control to 0.04 (Figure 8.14). Then $\pm 0.10/0.04 = \pm 2.5 \ \sigma$ variation is acceptable. Now 98.8% of our product meets the specification.

We next lower our σ to 0.03 and we get: $\pm 0.10/0.03 = \pm 3.34 \ \sigma$ varia-tion is acceptable. Then 99.93% of our product is acceptable.

Now if we happen to lose control of our process and our average value varies, we may still have problems of tolerance. See Figure 8.14 where these cases are plotted. Also note that loss of control of either σ or the average is very dangerous.

If one plots the averages and the sigmas, the losses (out of tolerance) can be read off the curve for both ends and then summated.

EXAMPLE E8-23.   The available commercial spread should preferably be somewhat larger than the $\pm 3\sigma$ limits of production.

In Figure 8.15 the plot shows the desirable situation where temporary loss of control of either the average or the σ does not result in rejections (or very low ones).

EXAMPLE E8-24.   Now suppose there was a gradual shifting of the average, say, due to the wearing by a part. This would result in a shift of the curve until a rejection limit was attained. Starting with a curve placed mid-way within the tolerance would not be the most economic answer. One would preferably start nearer one limit so that the wear would allow a shift toward the other limit. Figure 8.16 illustrates the problem.

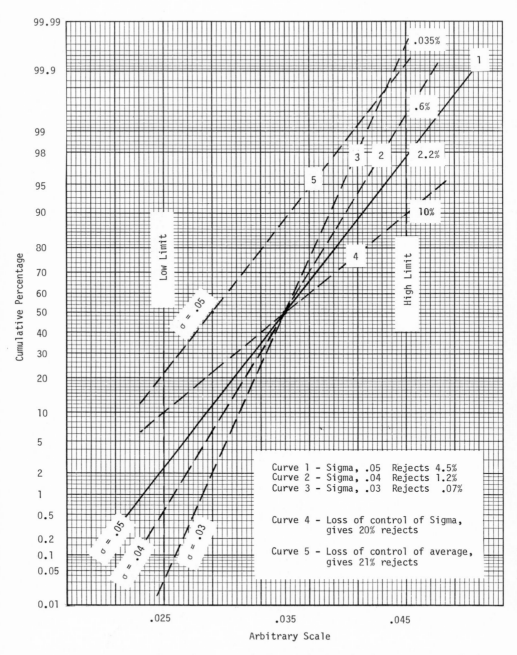

Figure 8.14 Effect of sigma on rejections at both ends of the curve. Total rejects are from both ends of curve.

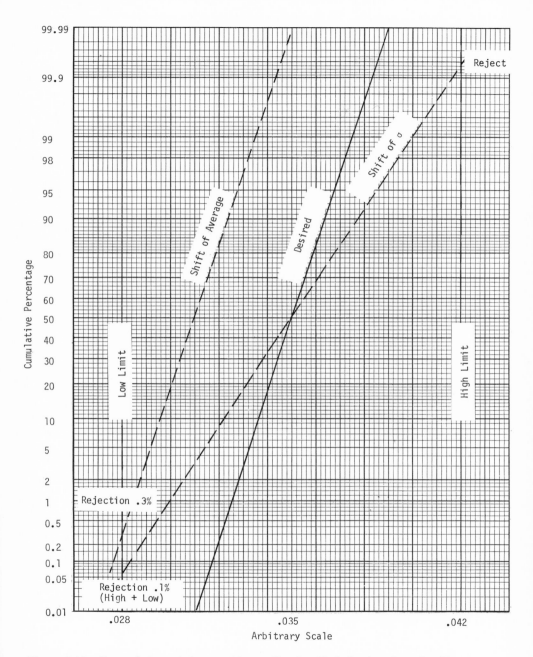

Figure 8.15 Case of sigma balance with specification limit for low loss. See Example E8-23.

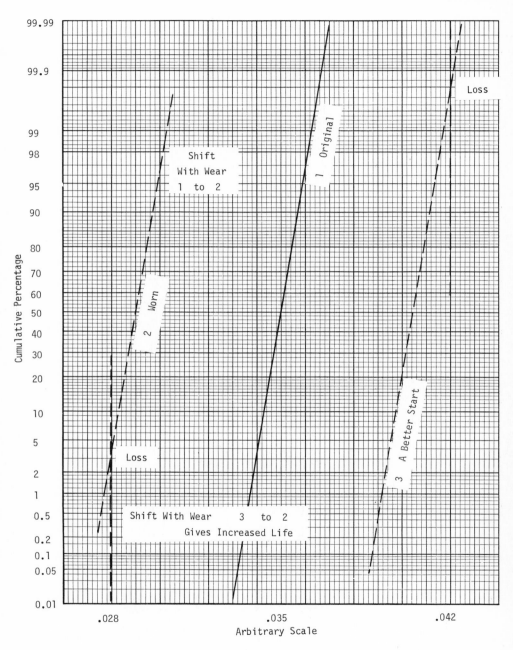

Figure 8.16 An illustration of wear on defects related to starting position. See Example E8-24.

The illustration of tubing manufacture with three distinct variables and precisions (sigmas) shown in Figure 8.3, Section 8.5, is a good example of a case where high precision may allow considerable shift to the left or right of the line without going out of tolerance any more than that normally obtained under low precision and no shift in average value.

EXAMPLE E8-25.   But there is always some tolerance of manufacture which may be unknown. The possible limits of the average and the σ must be recognized from a limited sample. The lines plotted in Figures 8.14, 8.15 and 8.16 then become areas and are fan shaped like those of Figure 8.13.

EXAMPLE E8-26.   These fan-shaped curves give an indication of quality as follows (Figure 9.2):

The central curve (lot B) tells us that about 8% of the glass (Figure 9.2) has a strength of 18,000 psi, or lower. Now, due to the fan-shaped uncertainty limits, this figure may run as low as 0.5% or as high as 27%. These are the intercepts of the fan curves with 18,000 abscissas (as indicated by arrows in Figure 9.2).

EXAMPLE E8-27.   In the case of a tolerance limit and a tolerance of inspection device, we have no clear-cut break of "go-no-go." As an example, if the production averages 0.060 thickness with a sigma of 0.003, we get the curve shown in Figure 8.17. Here we note that 0.38% of the product is below 0.052 value. If we arbitrarily set an inspection device at this 0.052 value, and the device has a sigma of 0.001, then the other line of Figure 8.17 is obtained for this device. Reading on the left scale, the device will reject $99\frac{1}{2}\%$ of samples below 0.050. This is a group average of these samples between 0.049 and 0.050. We will have a false acceptance of $\frac{1}{2}\%$ of the samples. If we wished to establish a 0.1% level, then the thickness would be 0.0506 as read on the arrows line of Figure 8.17. Following the arrows up to the inspection device line, we see that 91% would be rejected, and 9% accepted.

By blocking specific thickness regions together, we can determine acceptance or rejection. Some good is passed and some poor is accepted. Thus we get false acceptance and false rejections. This is expressed in Table 8.13. This more accurate calculation gives the following: of the 0.38% below 0.052, we accept 0.08, reject 0.30, while at the same time rejecting 0.28 above this level.

The *effects of resetting the inspection device curve* B right or left can be so calculated.

EXAMPLE E8-28.   An inspection procedure must not have a tolerance that is too large in relation to the desired limits of production. A good illustration of such a problem is that of density control of glass in a melting tank. R. D. Duff (Ref. 15) and L. G. Gehring (Ref. 18) have reported data on this topic. A sink-float technique is used whereby the temperature of the match-

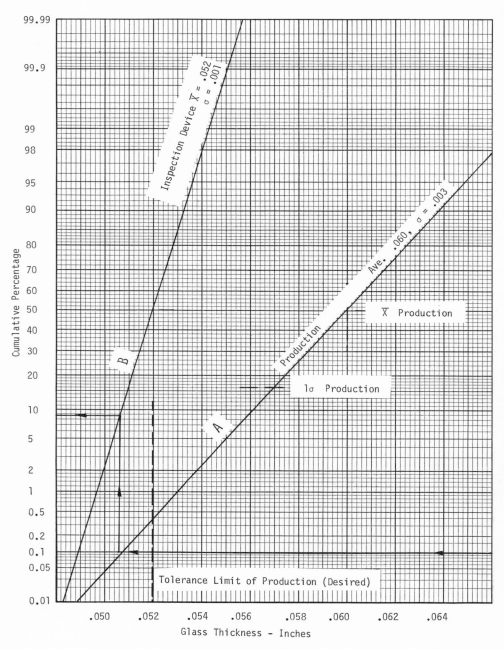

Figure 8.17 Relation of tolerance limit of production and rejections given by an inspection device. See Example E8-27.

Table 8.13 False Acceptance and False Rejection

Range	% in Group	% Rejected	Rejects	Accepts
Below 0.050	0.04	99.5	0.040	0.0002
0.050–0.051	0.09	94.0	0.085	0.0054
0.051–0.052	0.25	70.0	0.175	0.075
0.052–0.053	0.62	30.0	0.186	0.414
0.053–0.054	1.20	6.0	0.072	1.130
0.054–0.055	2.80	0.6	0.017	2.780
0.055–0.056	4.30	0.02	0.001	4.300
Totals 1st 3 items		Correct rejects	0.300	—
Below 0.052		False accepts	—	0.081
Totals last 4 items		False rejects	0.276	—
Above 0.052		Correct accepts	—	8.62
		Total rejects	0.576	

ing liquid is precisely controlled. This liquid changes 0.0019 in density for each 1°C change in its temperature. Combining the two workers' results, we see the following, given in density units:

first data, improperly trained technician $\quad \sigma = 0.0014$
after first training period $\quad \sigma = 0.0011$
after further training $\quad \sigma = 0.0008{-}0.0007$
averaging small subgroups $\quad \sigma = 0.0004{-}0.0003$
multiple repeat determinations of expert
   technician $\quad \sigma = 0.0001$
      to $\quad \sigma = 0.0002$
   with average $\quad \sigma = 0.00014$

It is desirable to hold the subgroup range of not more than 0.0012 density. On this basis the procedure proved useful. See Example E9-10 for related study.

8.12 THE RANGE (R) OF OBSERVATIONS. QUICK, EASY, DIRTY PRELIMINARY ANALYSIS OF EXTENSIVE DATA

The range (R) is the difference between the smallest and the largest value in a set of observations. It is often used where there are large amounts of data divided into cells each of which may be limited to 10 (or 15) values. The method is less burdensome, from a calculation viewpoint, and may be justified where speed is of greater importance than the other precise, or recognized, techniques. It is greatly affected by a single widespread value: the effect of this widespread value on σ would be much less. On this basis alone, we suggest that R be used on sample sizes from 3 to 10, occasionally even 15 to 20, never more.

EXAMPLE E8-29.   Take 4 subgroups, each of 5 observations, from a population:

$$R_1 = 1.073 \qquad R_2 = 1.433 \qquad R_3 = 4.313 \qquad R_4 = 3.738$$

$$\overline{R} = \frac{\sum(R_1 + R_2 \cdots + R_n)}{n} = \frac{10.557}{4} = 2.639$$

and compute, using the figures given in Table 8.14, where the sample size is multiplied by the number of lots to get the total number of samples and where we multiply R by the value of the multiplier in the table to estimate sigma.

Table 8.14   Using Range (R) Value to Estimate σ Versus Sample Size, from Table 8b (1), Ref. 13*

Sample Size (*lot size*)	Multiplier to Estimate σ from R	Variance (to be multiplied by σ^2)	Efficiency
2	0.886	0.571	1.000
3	0.591	0.275	0.992
4	0.486	0.183	0.975
5	0.430	0.138	0.955
6	0.394	0.112	0.933
7	0.370	0.095	0.911
8	0.351	0.083	0.890
9	0.337	0.074	0.869
10	0.325	0.067	0.850
15	0.288	0.047	0.766
20	0.268	0.038	0.700

Thus the multiplier from Table 8.14 for 5 observations is 0.430 and therefore

$$\sigma = (0.430)(2.639) = 1.13$$

The efficiency is the ratio of number of samples required for σ calculation versus the number of samples required for R calculation. Thus,

*From W. J. Dixon and F. J. Massey, Jr., *Introduction to Statistical Analysis*, 2nd ed., McGraw-Hill, 1957, by permission of McGraw-Hill Book Company. Copyright 1957 by McGraw-Hill Book Company.

when efficiency (5 groups) = 0.955

$$\frac{\text{samples for } \sigma}{\text{samples for } R} = 0.955$$

When we used 20 samples for R

$$\frac{n \text{ for } \sigma}{20} = 0.955$$

$$n \text{ for } \sigma = 19 \text{ approximately}$$

The confidence interval for sigma can be estimated from the equations (Ref. 13)

$$\text{low value} = \frac{(\text{multiplier})*(R)}{1 + 1.96\,[(\text{variance})^*/(\text{total number samples})**]^{\frac{1}{2}}}$$

$$\text{high value} = \frac{(\text{multiplier})*(R)}{1 - 1.96\,[(\text{variance})*/(\text{total number samples})**]^{\frac{1}{2}}}$$

when * = values from table for small lot size and ** = total number of lots multiplied by lot size. The 1.96 comes from 95% confidence limits (two-tailed). For above case:

$$\text{low value } \sigma = \frac{(0.430)2.639}{1 + 1.96(0.138/20)^{\frac{1}{2}}} = \frac{1.13}{1 + 1.96(0.0069)^{\frac{1}{2}}}$$

$$\text{low value } \sigma = \frac{1.13}{1 + 1.96(0.083)} = \frac{1.13}{1.16} = 0.97$$

$$\text{high value } \sigma = \frac{1.13}{0.84} = 1.35$$

Range for σ is from 0.97 to 1.35. The 95% confidence limits above are based on 0.95 two-tailed normal distribution (value 1.96). See Table 9.3 for other values.

Dixon and Massey (Ref. 13), text pages 272–278, and table pages 404–412, give other so-called "quick" or "microstatistics" for substitute t ratio, F ratio, etc. These authors suggest that such an approach may be quite adequate for preliminary studies or slide rule rapid computations, particularly where a survey of a large data field is needed to outline the principal area of study, or to determine what portion of the data may require extensive analysis.

9

COMPARISON OF TWO SETS
OF DATA FOR
SIGNIFICANT DIFFERENCES

"False facts are highly injurious —
for they often endure long."

Darwin

9.1 THE SIMPLE CASE

When the averages of two large groups differ greatly and the standard deviations are small, there is no problem. That is, the data do not significantly overlap at all.

EXAMPLE E9-1. In the data for Table 9.1 the total ranges do not overlap. There is no question of average or range. See Figure 9.1.

Table 9.1 Averages of Two Groups with No Significant Overlap

	Average	σ	3σ	Total Range of Uncertainty (0.997 Probability)
Set 1	180	10	30	150–210
Set 2	120	10	30	90–150

9.2 THE GRAPHICAL PROCEDURE

By the use of probability paper, we may determine significant differences. As an illustration, take the following example:

EXAMPLE E9-2. The transverse strengths of glass rods were studied in two lots, *A* and *B*. Figure 9.2 gives the probability plots of 19 samples of

109

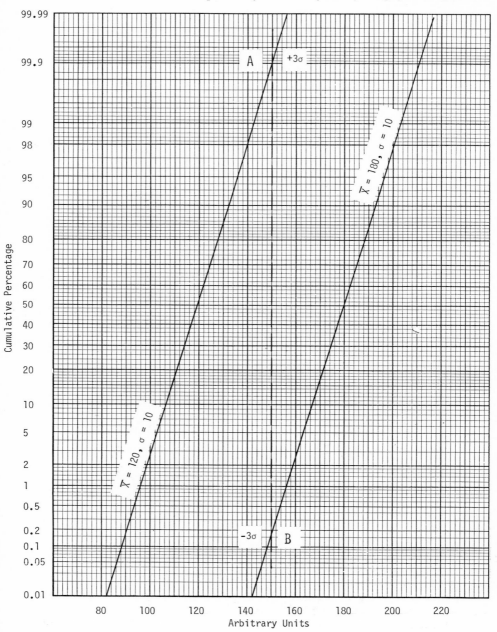

Figure 9.1   The case of no overlap of 3σ limits; A and B values are identical. See Example E9-1.

Table 9.2 Strength of Two Groups of Glasses

Lot	Average	σ	Number (n)
Lot B	24.5	4.7	19
Lot A	20.0	3.2	10

lot B; 20.9, 25.1, 30.6, 31.3, 15.2, 22.6, 20.3, 23.9, 22.8, 25.9, 18.8, 23.9, 20.0, 30.6, 31.7, 31.5, 24.2, 18.5, 26.7, in order of test (all times 1,000 psi). In a like fashion 10 samples of lot A gave 22.3, 23.5, 19.0, 17.7, 14.5, 15.1, 21.4, 20.8, 19.0, 23.9. Plot as directed earlier (Figure 9.2), using the data as in Table 9.2. The findings for lot A, when Range $\bar{X}' = \bar{X} \pm K\sigma$ (from Section 8.8), on 95% confidence, $n = 10$, $K = 0.715$ (Table 8.9) are

$$\text{Range } \bar{X}' = 20.0 \pm (0.715)(3.2) = 20.0 \pm 2.29 = 17.7 \text{ to } 22.3$$

Range $\sigma' = A_L \sigma$ to $A_U \sigma$ (from Section 8.9); $A_L = 0.688$ and $A_U = 1.83$ for $n = 10$ and 95% confidence (Table 8.12).

$$\text{Range } \sigma' = (0.688)(3.2) \text{ to } (1.83)(3.2) = 2.2 \text{ to } 5.9$$

Lot B, when Range $\bar{X}' = \bar{X} \pm K\sigma$, on 95% confidence, $n = 19$, $K = 0.482$, gives

$$\text{Range } \bar{X}' = 24.5 \pm (0.482)(4.7) = 24.5 \pm 2.27 = 22.2 \text{ to } 26.8$$

Range $\sigma' = A_L\sigma$ to $A_U\sigma$; $A_L = 0.755$ and $A_U = 1.48$ for $n = 19$ (Table 8.12).

$$\text{Range } \sigma' = (0.755)(4.7) \text{ to } (1.48)(4.7) = 3.55 \text{ to } 6.96 = 3.6 \text{ to } 7.0$$

Plotting as instructed under Section 8.10, we get the fans of uncertainty as shown in Figure 9.2. There is a very slight overlap of the averages, that is,

Range $\bar{X}'$ for A = 17.7 to 22.3   Range $\bar{X}'$ for B = 22.2 to 26.8

There is overlap (crosshatched) of the two fans, based on the estimated range of the σ'. What one really needs here is a larger sample for Lot A to narrow the fan more closely. In this case, we believe that there is significant difference shown, but the ultimate test would be the t test, to be later described, which does definitely show significant difference of Lot A and Lot B at something over 99%.

So the t test should be run on these data. It is easier to do than what we have already done above.

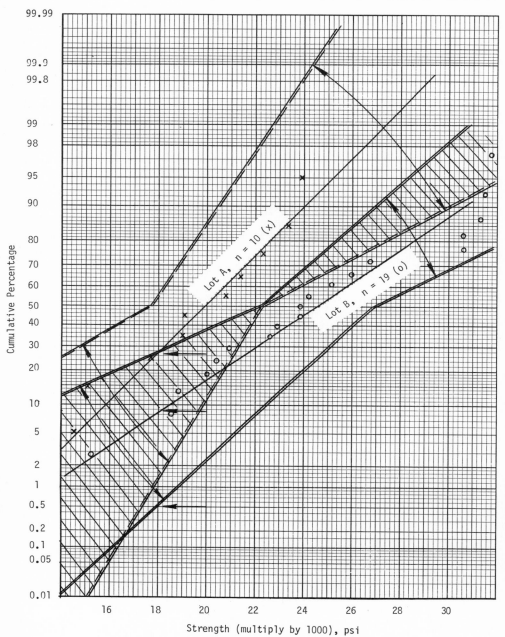

Figure 9.2   Transverse strengths of two lots of glass. See Examples E8-26 and E9-2.

9.3 DETERMINATION OF WHETHER THERE ARE SUFFICIENT DATA FOR ESTABLISHMENT OF SIGNIFICANT DIFFERENCES

ASSUMING CONFIDENCE OF 95% — USUALLY SATISFYING A PRELIMINARY INVESTIGATION.   A convenient chart relating number of samples required to establish significant differences between two sets of data is given in Figure 9.3. It is based on a relationship between the sum of the standard deviations and the difference in the averages. With these two figures, we can then read off how many samples would be required in each group. This chart is based on a 95% confidence envelope, and on equal sample sizes in two groups.

EXAMPLE E9-3.   Suppose that in the *strength test* of Section 9.2, Table 9.2, we had 20 samples in each group. Then,

Sum of sigmas, $4.7 + 3.2 = 7.9$
Difference of averages. $24.5 - 20.0 = 4.5$.
Chart, Figure 9.3, reads 14 samples from each group required.

Incidentally, the chart is useful for cases where the difference in the average is greater than 9. Simply multiply both parameters $(X_1 - X_2)$ and $(\sigma_1 + \sigma_2)$ by 10, *without* changing the value of n, the number of tests.

EXAMPLE E9-4.   Suppose two groups of ten products each had *dimensional* averages, each of which had a σ of 0.003. What must be the difference in average to be significant?

$$\sigma_1 + \sigma_2 = 0.003 + 0.003 = 0.006$$

Read off abscissas of 6, equivalent 0.006. Read off ordinate of 10. Then get $(X_1 - X_2)$ of 3.5, equivalent 0.0035.

Difference less than 0.0035 would require larger sample size than this average of ten products.

In a like fashion, the necessary σ values can be estimated if the average differences and number of samples are known.

CHART BASED ON CONFIDENCE FROM 50% TO 99% — USUALLY SATISFYING A MAJOR EFFORT.   Sometimes we wish to develop a concept of the variation of the confidence based on sample size. In 99 sets out of 100 sets of data, Figure 9.4 gives the sample size (n) based on 50% to 90% confidence for the detection of a difference between two averages.

How to Use Figure 9.3 to Determine Sample Size

Figure 9.3 is used to determine how many samples are necessary for establishment of significant difference for *variables*, and assuming 95% confidence.

Given the average and sigma for each of two groups of data of *equal size;* are there enough data, or is the sample size big enough for a determination of significant differences between the two sets of data?

Summate the sigmas (add them together).

Determine difference in average (subtract them).

Read off chart of Figure 9.3 from bottom up for summation of sigmas, to the curve for difference of averages, and read to left for sample size.

For example:

        average 58, sigma 3,   average 49, sigma 6
     summation of sigmas is 9   difference in average is 9

Figure 9.3 reads off as 5 samples required in each group.

Another example:

     average 0.048, sigma 0.007   average 0.044, sigma 0.008
Read off 0.015 and 0.004 and get sample number of 56 required.

Still another example:

       average 700, sigma 70   average 600, sigma 60
               difference 100, sum 130
Read off 9 or 10 samples required in *each* group.

Note: The numbers on abscissas and on curves may be multiplied by any number, say 10 or 100, if both are so treated. *Do not multiply the sample number.*

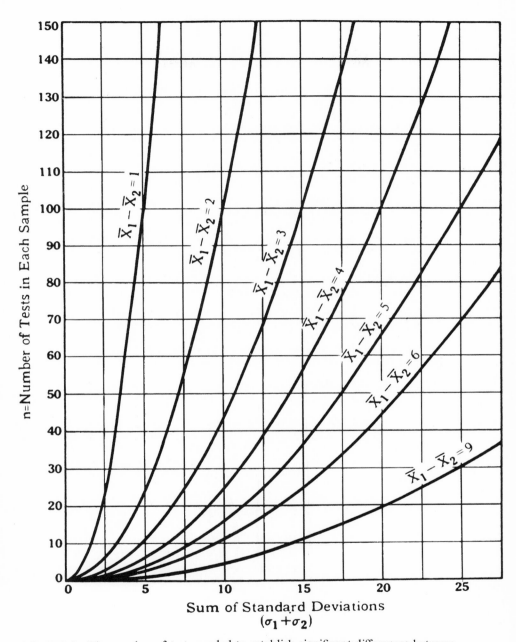

Figure 9.3 The number of tests needed to establish significant differences between two sets of data, based on 95% confidence. From Reference 21, L. R. Hill and P. L. Schmidt, "Graphical Statistics — An Engineering Approach," *Westinghouse Engr.* March 1950 and May 1950.

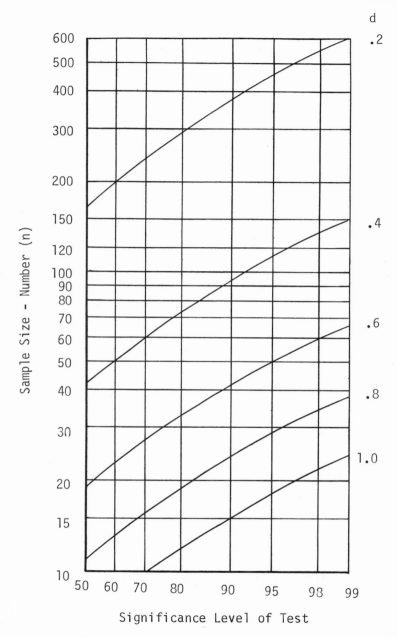

Figure 9.4 Sample size required for detection of difference between averages. Basis: 99 sets out of 100 sets. From Reference 26, M. G. Natrella, Experimental Statistics, *Natl. Bur. Std. Handbook*, 91, Aug 1, 1963.

Figure 9.4 (based on Table A-8 of Ref. 26) uses the following nomenclature:

Case 1: To detect if a new product differs from a standard product. If n = sample size, m_0 = average of an accepted product, and m = average of a new product, then

$$d = \frac{m - m_0}{\sigma}$$

We calculate d, and read off n directly if we know the value of σ. If we do not know the value of σ, and must estimate it from our sample, then we must add 4 units to our n value as a correction.

Case 2: To detect if two product averages are significantly different. If n = sample size, m_A = average for product A, m_B = average for product B, and $\sigma_A = \sigma_B$, then

$$d = \frac{m_A - m_B}{(\sigma_A^2 + \sigma_B^2)^{\frac{1}{2}}}$$

We calculate d and read off n, then add a correction factor of 2 to n.

In these cases the d value, known as the standardized difference, tells us, with a probability — listed on the graph — whether the averages are significantly different. *The sample sizes for these comparisons must be the same.*

EXAMPLE E9-5. Recalculating Example E9-2, assuming σ for each value is 4, use Case 2.

$$d = \frac{(24.5 - 20.0)}{(4^2 + 4^2)^{\frac{1}{2}}} = \frac{4.5}{32^{\frac{1}{2}}} = 0.8$$

Read Figure 9.4 for 0.8 to get $n = 29$; n corrected $= 29 + 2 = 31$.

EXAMPLE E9-6. If we accept the value of average 20, sigma 4, as a standard product value — obviously better known statistically because of the acceptance — then we use Case 1.

$$d = \frac{24.5 - 20.0}{4} = \frac{4.5}{4} = 1.1$$

Read $n = 15$ (an extrapolated value beyond our chart); no correction to this value n. This computation is a good check for Figure 9.3.

EXAMPLE E9-7. For the data of example accompanying the explanation for use of Figure 9.3, we use Case 1, recognizing that $m_0 = 0.044$, $\sigma = 0.0075$, and

$$d = \frac{(0.048 - 0.044)}{0.0075} = 0.53$$

Figure 9.4 reads off about $n = 65$; no correction.

EXAMPLE E9-8.   For the data $m = 280$, $m_0 = 275$, and $\sigma = 7$, we get

$$d = \frac{m - m_0}{\sigma} = \frac{280 - 275}{7} = 0.7$$

Read off from Figure 9.4, $n = 37$; no correction. Figure 9.3 reads $n = 34$.

EXAMPLE E9-9.   For the data of Example E9-7, we assume both are experimental lots, with $\sigma = 0.0075$. Use Case 2.

$$d = \frac{(0.048 - 0.044)}{[(0.0075)^2 + (0.0075)^2]^{\frac{1}{2}}} = \frac{0.004}{0.0013} = 0.35$$

Read off Figure 9.4 that n is about 160 (interpolation required).

Our main conclusion: Really good comparisons of qualities of two experimental products require large sample sizes.

Note: Figure 9.4 is calculated as follows: $Z =$ the distance from the population mean *in units of the standard deviation*, $n = (Z_{0.995} + Z_P)^2/d^2$ where subscript P refers to the significance level of the test. For our tabulation $Z_{0.995} = 2.576$ and Z_P values are

0.00   for 0.50 probability
0.52   for 0.70 probability
1.28   for 0.90 probability
1.64   for 0.95 probability
2.05   for 0.98 probability
2.326 for 0.99 probability

9.4   THE t TEST

The simple procedure given in Section 9.1 is useful where no overlap of the expected averages is found. However, there may still be significant differences between two sets of data even with overlap of these "limits of uncertainty."

The recognized statistical test is known as the t test and involves the calculation of the standard deviation of the difference of these averages. We then define t as:

$$t = \frac{\text{difference between the averages}}{\text{standard deviation of the difference}}$$

Now if t exceeds certain values defined in a table, and based on the number of samples used for each average, we can say that there is evidence of a difference between the two sets of data.

The *t* test is based on the assumption that both populations are normally distributed with perhaps different means (X'_A differs from X'_B) but similar sigmas ($\sigma'_A = \sigma'_B$).

The practical engineer need not be too concerned if he uses 10 or more samples from each group, since fairly large departures from normality can be tolerated.

For differences in sigma, the *F* test described in Section 10.2 should be used. We may then fall back to the *t* test. We may have to reestimate our sample size, and get the proper amount of data.

Our *t* table is based on a two-tailed (two-sided) concept — all distribution curves have a high and a low tail. This is the more generally useful table since we are usually interested in *whether a difference* in average exists. If we want to determine *if an average is larger* than another average, we must use a one-tailed concept. Such *t* values are read off using *P* equal to half those listed in Table 9.3.

WHERE AVERAGES AND SIGMAS ARE KNOWN. It is most convenient to use the *t* test if you have the average and the sigma for each average for the two separate groups which you desire to compare. In this case, the "standard deviation of the difference," known as σ_D, is calculated by the formula:

$$\sigma_D = \left[\frac{(\sigma_n)^2}{n} + \frac{(\sigma_m)^2}{m} \right]^{\frac{1}{2}}$$

where *n* and *m* represent the two groups, as well as the number of samples.

EXAMPLE E9-10. In a glass study, by R. D. Duff (Ref. 15), of the change in density from lehr-annealing to laboratory annealing, Mr. Duff plotted the changes for groups equivalent to a No. 1 and No. 2 strain disk. Sufficient samples (290 and 240 respectively) were used to give a very good probability plot (Figure 9.5). The group lots were as shown in Table 9.4, where the density change is as measured by temperature change of density liquid. Calculate:

$$\sigma_D = \left[\frac{(\sigma_n)^2}{n} + \frac{(\sigma_m)^2}{m} \right]^{\frac{1}{2}}$$

$$= \left[\frac{(0.20)^2}{290} + \frac{(0.21)^2}{240} \right]^{\frac{1}{2}} = (0.000138 + 0.000184)^{\frac{1}{2}} = 0.017$$

$$t = \frac{0.70 - 0.56}{0.017} = \frac{0.14}{0.017} = 8.2$$

Formulas for Usage of *t* **Table 9.3**

The *t* table, Table 9.3, gives values of the ratio

$$\frac{\text{difference between the averages}}{\text{standard deviation of the difference}}$$

For variables only

$$t = \frac{\bar{X}_n - \bar{X}_m}{\sigma_D}$$

where *n* is the number in the *n* group and where *m* is the number in the *m* group. Calculate σ_S, the pooled standard deviation of the two groups taken as one group

$$\sigma_D = \frac{\sigma_S}{[nm/(n + m)]^{\frac{1}{2}}} \quad \text{or} \quad \sigma_D = \sigma_S \left(\frac{n + m}{nm}\right)^{\frac{1}{2}}$$

$$\sigma_S = \left[\frac{(\sigma_n)^2(n - 1) + (\sigma_m)^2(m - 1)}{n + m - 2}\right]^{\frac{1}{2}}$$

$$\sigma_D = \left(\frac{\sigma_n^2}{n} + \frac{\sigma_m^2}{m}\right)^{\frac{1}{2}}$$

$$t = \frac{\bar{X}_n - \bar{X}_m}{\sigma_D}$$

Degrees Freedom (for *t* table only) = total number of samples in both groups minus 2, or

$$df = (n - 1) + (m - 1)$$

(For discussion of df see Section 9.5.)

The confidence of the decision is based on the *t* table, as follows:

0.01 probability exceeded, very positive confidence
0.05 probability exceeded, confidence is acceptable
0.10 probability exceeded, confidence has some question

Table 9.3   Critical Values of *t* Based on Two Tails of Curve (from Ref. 16)

df	0.50	0.40	0.30	0.20	0.10	0.05	0.02	0.01	0.001
					Probability				
1	1.000	1.376	1.963	3.078	6.314	12.706	31.82	63.66	636.62
2	0.816	1.061	1.386	1.886	2.920	4.303	6.97	9.93	31.60
3	0.765	0.978	1.250	1.638	2.353	3.182	4.54	5.84	12.94
4	0.741	0.941	1.190	1.533	2.132	2.776	3.75	4.60	8.61
5	0.727	0.920	1.156	1.476	2.015	2.571	3.37	4.03	6.86
6	0.718	0.906	1.134	1.440	1.943	2.447	3.14	3.71	5.96
7	0.711	0.896	1.119	1.415	1.895	2.365	3.00	3.50	5.41
8	0.706	0.889	1.108	1.397	1.860	2.306	2.90	3.36	5.04
9	0.703	0.883	1.100	1.383	1.833	2.262	2.82	3.25	4.78
10	0.700	0.879	1.093	1.372	1.812	2.228	2.76	3.17	4.59
11	0.697	0.876	1.088	1.363	1.796	2.201	2.72	3.11	4.44
12	0.695	0.873	1.083	1.356	1.782	2.179	2.68	3.06	4.32
13	0.694	0.870	1.079	1.350	1.771	2.160	2.65	3.01	4.22
14	0.692	0.868	1.076	1.345	1.761	2.145	2.62	2.98	4.14
15	0.691	0.866	1.074	1.341	1.753	2.131	2.60	2.95	4.07
16	0.690	0.865	1.071	1.337	1.746	2.120	2.58	2.92	4.02
17	0.689	0.863	1.069	1.333	1.740	2.110	2.57	2.90	3.97
18	0.688	0.862	1.067	1.330	1.734	2.101	2.55	2.88	3.92
19	0.688	0.861	1.066	1.328	1.729	2.093	2.54	2.86	3.88
20	0.687	0.860	1.064	1.325	1.725	2.086	2.53	2.85	3.85
21	0.686	0.859	1.063	1.323	1.721	2.080	2.52	2.83	3.82
22	0.686	0.858	1.061	1.321	1.717	2.074	2.51	2.82	3.79
23	0.685	0.858	1.060	1.319	1.714	2.069	2.50	2.81	3.77
24	0.685	0.857	1.059	1.318	1.711	2.064	2.49	2.80	3.75
25	0.684	0.856	1.058	1.316	1.708	2.060	2.48	2.79	3.73
26	0.684	0.856	1.058	1.315	1.706	2.056	2.48	2.78	3.71
27	0.684	0.855	1.057	1.314	1.703	2.052	2.47	2.77	3.69
28	0.683	0.855	1.056	1.313	1.701	2.048	2.47	2.76	3.67
29	0.683	0.854	1.055	1.311	1.699	2.045	2.46	2.76	3.66
30	0.683	0.854	1.055	1.310	1.697	2.042	2.46	2.75	3.65
40	0.681	0.851	1.050	1.303	1.684	2.021	2.42	2.70	3.55
60	0.679	0.848	1.046	1.296	1.671	2.000	2.39	2.66	3.46
120	0.677	0.845	1.041	1.289	1.658	1.980	2.36	2.62	3.37
∞	0.674	0.842	1.036	1.282	1.645	1.960	2.33	2.58	3.29

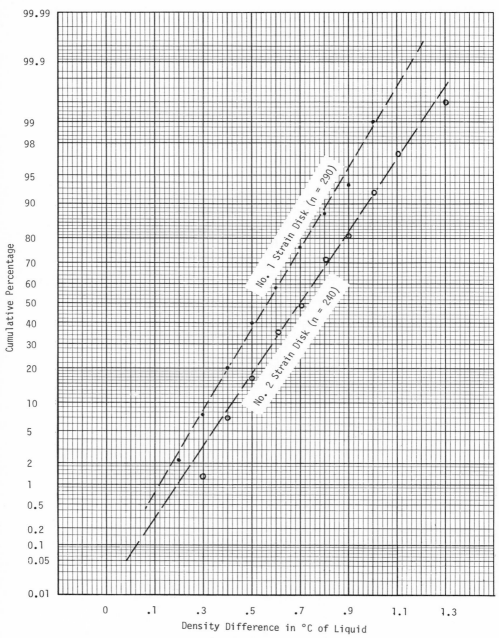

Figure 9.5 Glass density difference as a result of reannealing from no. 1 or no. 2 strain disks, when 1°C = 0.0019 density. See Example E9-10. From Reference 15, R. D. Duff, *J. Am. Ceram. Soc.* 30:12, 1947, by permission of the American Ceramic Society. Copyright 1947 by the American Ceramic Society.

Table 9.4 Group Lots for Density of Glass (from Ref. 15)*

	Density Change	σ	No. of Samples
No. 1	0.56	0.20	290
No. 2	0.70	0.21	240

Table 9.3 gives the critical values for *t* based on degrees of freedom (df). We will describe this term later (Section 9.5), but it is sufficient to say here that 290 values give 289 df and 240 values give 239 df. So our total df equals 289 plus 239 = 528, or, for purposes of Table 9.3, df = ∞ (infinity).

The Table 9.3 *t* value for ∞ df is 3.29 for probability of 0.001. Our *t* calculated at 8.2 far exceeds 3.29 and therefore the difference between the two sets of data is very significant, since it could occur only 1 time in 1000 times if the two populations were equal.

We calculated above the standard deviation of the difference, σ_D, as 0.017. Now the actual difference in averages was:

$$0.70 - 0.56 = 0.14$$

We know that for a normal distribution, a difference as large as $3\sigma_D$ might arise due to chance (997 times out of 1,000).

$$3\sigma_D = 3 \times 0.017 = 0.051$$

Since 0.051 is far less than 0.14, we know the differences are real and not due to sampling or sample size or unusual variation.

EXAMPLE E9-11.   In a study of the time interval in a measured process, we ran 25 tests under each of 4 conditions. The times were taken electronically, with the data given in Table 9.5.

Table 9.5 Time Interval, Seconds, for a Measured Process

Condition	Time in Seconds			
	Min.	Ave.	Max.	Sigma
(1)	0.441	0.479	0.527	0.028
(2)	0.463	0.472	0.483	0.006
(3)	0.457	0.518	0.539	0.019
(4)	0.467	0.477	0.486	0.007

*From R. D. Duff, *J. Am. Ceram. Soc.*, 30:12, 1947, by permission of the American Ceramic Society. Copyright 1947 by the American Ceramic Society.

Table 9.6 Calculations for Example E9-11

Groups	Difference in Averages	σ of Difference	t
1 & 2	0.007	0.006	1+
3 & 4	0.041	0.005	8+
1 & 3	0.039	0.007	5+
2 & 4	0.005	0.002	2+

Using the formula (sigma of difference):

$$\sigma_D = \left[\frac{(\sigma_n)^2}{n} + \frac{(\sigma_m)^2}{m}\right]^{\frac{1}{2}}$$

the standard deviation of the differences of the four combinations was calculated as in Table 9.6, where df = (25 − 1) + (25 − 1) = 48; t (Table 9.3) for df = 3.5.

The ts were calculated as above and significance established for comparisons of 3 and 4 as well as 1 and 3. Comparisons of 1 and 2 as well as 2 and 4 were not significant. Therefore, the following conclusions were justified:

1. Condition (4) is definitely superior (the time is shorter) than condition (3).
2. Condition (3) is definitely poorer (the time is longer) than condition (1).
3. There is no significant difference between conditions (1) and (2).
4. There is no significant difference between conditions (2) and (4).

EXAMPLE E9-12.   We have three sets of data on strength of glass. The results are plotted on probability paper in Figure 9.6. The general results are:

Group 1   av 0.93   σ 0.13
Group 2   av 0.72   σ 0.17
Group 3   av 0.67   σ 0.14

On the basis that the data were a result of at least 100 tests, we can calculate t.

Comparing Groups 1 and 2

$$\sigma_D = \left(\frac{0.13^2}{100} + \frac{0.17^2}{100}\right)^{\frac{1}{2}}$$

$$\sigma_D = 0.021$$

$$t = \frac{0.93 - 0.72}{0.021} = \frac{0.21}{0.021} = 10$$

t Table at ∞   and   0.001 = 3.29

Difference is highly significant.

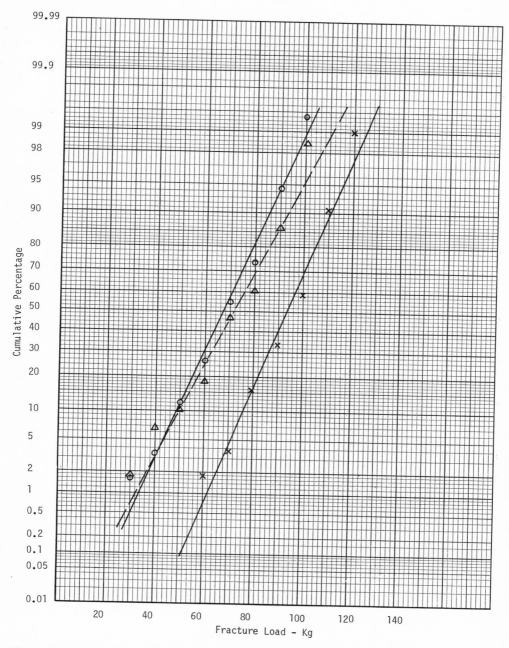

Figure 9.6 Three sets of data on strength of glass. Group 1 is represented by ✕, group 2 by △, and group 3 by ◯. See Example E9-12.

Comparing Groups 2 and 3

$$\sigma_D = \left(\frac{0.17^2}{100} + \frac{0.14^2}{100}\right)^{\frac{1}{2}}$$

$$\sigma_D = 0.022$$

$$t = \frac{0.72 - 0.67}{.022} = \frac{0.05}{0.022} = 2+$$

Difference is not significant.

EXAMPLE E9-13. Take the example (Table 9.7) of two analysts running % C on a chemical, ephedrine. (Example from Ref. 36.)

Table 9.7   Two Analysts' Results for Precision and Accuracy*

	Analyst I	*Analyst II*
Number of analyses (n)	6	4
Specific results	59.09	59.51
	59.17	59.75
	59.27	59.61
	59.13	59.60
	59.10	
	59.14	
Average	59.15	59.62
Degrees of freedom ($n - 1$)	5	3
Squares of difference from average	0.06^2	0.11^2
	0.02^2	0.13^2
	0.12^2	0.01^2
	0.02^2	0.02^2
	0.05^2	
	0.01^2	
Sum of squares	0.0214	0.0295

$$\sigma_S = \left(\frac{0.0214 + 0.0295}{5 + 3}\right)^{\frac{1}{2}} = 0.0797$$

Note here that we have not bothered to calculate σ_I or σ_{II} for the

*From W. J. Youden, *Statistical Methods for Chemists*, Wiley, 1951, by permission of John Wiley & Sons, Inc. Copyright 1951 by John Wiley & Sons, Inc.

analysts, but rather used the sum of squares directly. That is

$$\text{sum of squares} = (\sigma_I)^2(n - 1) \qquad \text{and} \qquad (\sigma_{II})^2(m - 1)$$

$$\sigma_D = \frac{0.0797}{[nm/(n + m)]^{\frac{1}{2}}} = \frac{0.0797}{[(6 \times 4)/(6 + 4)]^{\frac{1}{2}}}$$

$$= \frac{0.0797}{(2.4)^{\frac{1}{2}}}$$

$$\sigma_D = \frac{0.0797}{1.55} = 0.0514$$

$$t = \frac{59.62 - 59.15}{0.0514} = \frac{0.47}{0.0514} = 9.15$$

Critical values for t, 8 degrees of freedom are (from Table 9.3) given in Table 9.8. The $t = 9.15$ far exceeds any value for 8 df from the table. We can thus conclude that the observed difference in analytical results $(59.62 - 59.15)$ of 0.47% is not reconcilable with the zero difference expected if both analysts had the same bias. The difference in average is significant. There must be a reason. Look for it.

Table 9.8 Critical Values for t, 8 Degrees of Freedom

	Probability Level						
	0.50	0.40	0.30	0.20	0.10	0.05	0.01
8 df	0.706	0.889	1.108	1.397	1.860	2.306	3.355

The standard deviation of the average difference (as calculated in Example E9-13)

$$\sigma_D = 0.0514$$

allows a calculation based on an engineering probability of 0.95 as follows:

95% confidence limits
high limit $0.47 + 2.306 \, (0.0514) = 0.59$
low limit $0.47 - 2.306 \, (0.0514) = 0.35$

The figure 2.306 is from t table, 8 df and 95% confidence. Thus, t is really a multiplier for σ to give confidence ranges with any desired degree of certainty.

Thus, the conclusion is that the results of Analyst I may be expected to be low relative to those of Analyst II by an amount somewhere between 0.35 and 0.59% in 19 out of 20 cases.

Remembering that the confidence limits of an average may be predicted by $t\sigma/n^{\frac{1}{2}}$ (see Section 8.8), we must now calculate σ for each analyst.

Confidence Limits — Analyst I

$$\sigma = \left(\frac{\text{sum of squares}}{n-1}\right)^{\frac{1}{2}} = \left(\frac{0.0214}{5}\right)^{\frac{1}{2}} = 0.0654$$

Now

$$\frac{\sigma}{n^{\frac{1}{2}}} = \frac{0.0654}{6^{\frac{1}{2}}} = 0.0267$$

Confidence limits for $\bar{X}$ for Analyst I, according to equation

$$\text{Range of } \bar{X}' = \bar{X} \pm t\frac{\sigma}{n^{\frac{1}{2}}}$$

and therefore

$$59.15 - (0.0267)(2.571) = 59.08$$
$$59.15 + (0.0267)(2.571) = 59.22$$
$$\text{Range of } \bar{X}' = 59.08 \text{ to } 59.22$$

(Factor 2.571 is the t value for 5 df at 95% confidence, from Table 9.3.)

Confidence Limits — Analyst II

$$\sigma = \left(\frac{\text{Sum of Squares}}{n-1}\right)^{\frac{1}{2}} = \left(\frac{0.0295}{3}\right)^{\frac{1}{2}} = 0.0992 \qquad \frac{\sigma}{n^{\frac{1}{2}}} = \frac{0.0992}{4^{\frac{1}{2}}} = 0.0496$$

$$\bar{X}' = 59.62 \pm (0.0496)(3.182) \qquad \text{Range of } \bar{X}' = 59.46 \text{ to } 59.78$$

(Factor 3.182 from t table 9.3 for 3 df at 95% confidence.)

True content in the chemical was 59.55 so Analyst II bracketed the true result. Analyst I had some bias in procedure or an impure sample.

From the above example we see the usage for the t table and its possibility in calculation. It allows not only an estimation of average, but also a confidence limit for each average. This method for the latter calculation supplements those given in Section 8.8.

SUMMARY.   It is not necessary to know the individual standard deviations to make a calculation based on the t table.

The procedure is to lump all the measurements into a single group and determine the standard deviation of the group.

$$\sigma_S = \text{sigma of the two groups together}$$

Then, where n = number of group 1 and m = number of group 2,

$$\sigma_D = \frac{\sigma_S}{[nm/(n+m)]^{\frac{1}{2}}} \qquad \text{or} \qquad \sigma_D = \sigma_S\left(\frac{n+m}{nm}\right)^{\frac{1}{2}}$$

$$t = \frac{n_{av} - m_{av}}{\sigma_D} = \frac{X_n - X_m}{\sigma_D}$$

Then compare the above to Table 9.3 for *t* value for df $= (n + m) - 2$.

WHERE ONLY PAIRS OF RESULTS ARE KNOWN ON A NUMBER OF DIFFERENT SAMPLES — TO COMPARE SIGNIFICANCE OF ONE OF THE PAIRS VERSUS THE OTHER. Sometimes procedures are changed, and the old is compared with the new by each method on different samples. Do the procedures give significant differences?

EXAMPLE E9-14. Take 10 results of durability (mg NaOH titration) on Glasses A to J as in Table 9.9.

Table 9.9 Pairs of Results — Durability of Glasses

	Glass									
	A	B	C	D	E	F	G	H	I	J
1st result	23	5	26	11	14	27	8	19	36	23
2nd result	23	5	23	12	13	24	9	20	35	22
Difference	0	0	3	−1	1	3	−1	−1	1	1
Difference²	0	0	9	1	1	9	1	1	1	1

Summation difference $= 6$
Summation (difference)² $= 24$

Sigma (of differences) $= \sigma_D = \left[\frac{(24) - (6^2/10)}{10 - 1}\right]^{\frac{1}{2}} = 2.27^{\frac{1}{2}} = 1.51$

But this sigma depended upon 10 pairs of results, and so must be divided by $n^{\frac{1}{2}}$ or $10^{\frac{1}{2}}$ to correct for the 10 different glasses.

Now if the procedures (1st result and 2nd result) are identical, or are assumed identical, we know that the *expected difference* (E) should be zero; but the actual difference was 6 divided by the number of glasses or:

$$\bar{X} = \frac{6}{10} = 0.6$$

Therefore,

$$t = \frac{\bar{X} - E}{\sigma/n^{\frac{1}{2}}} = \frac{0.6 - 0}{1.51/n^{\frac{1}{2}}} = \frac{0.6}{0.48} = 1.25$$

Now we can evaluate the significance of this $t = 1.25$. The degrees of freedom $= 10 - 1 = 9$. In Table 9.3, the $t = 1.25$ falls between a probability of 0.30 and 0.20. If our original hypothesis were true — that there is no difference in procedure — we should expect a t value as large as 1.25 between 20% and 30% of the time under repeated testing. There is no evidence to doubt the validity of the original assumption. On the other hand, had our calculated t value been 2.26, then the t table shows this will happen only 5% of the time. In this case we have more reason to doubt the validity of the original assumption.

Thus the difference between 1st and 2d results is not significant.

EXAMPLE E9-15.   Another example based on two methods on different lots of raw material (paired procedures) is calculated in Table 9.10 (from Ref. 36).

Table 9.10 Data for Paired Procedures, Lead Content by Two Procedures*

	Percent Lead		$d = Difference$	Sum
Heat	Method X	Method Y	X − Y	X + Y
1	2.68	2.58	0.10	5.26
2	2.55	2.34	0.21	4.89
3	2.29	2.03	0.26	4.32
Total	7.52	6.95	0.57	14.47
Average	2.51	2.31	0.19	

Now

$\sigma_D =$ standard deviation of difference

$$\sigma_D{}^2 = \frac{1}{n-1}(d_1{}^2 + d_2{}^2 + d_3{}^2 - n\bar{d}^2)$$

$$\sigma_D{}^2 = \frac{1}{3-1}(0.10^2 + 0.21^2 + 0.26^2 - 3 \times 0.19^2)$$

$$\sigma_D{}^2 = 1/2 \,(0.0100 + 0.0441 + 0.0676 - 0.1083) = 0.0067$$

$$\sigma_D = (0.0067)^{\frac{1}{2}} = 0.082$$

$$t = \frac{\bar{d} - 0}{\sigma_D/n^{\frac{1}{2}}} = \frac{(\bar{d} - 0)n^{\frac{1}{2}}}{\sigma_D}$$

*From W. J. Youden, *Statistical Methods for Chemists*, Wiley, 1951, by permission of John Wiley & Sons, Inc. Copyright 1951 by John Wiley & Sons, Inc.

Where n equals the number of pairs, $n^{\frac{1}{2}}$ enters the picture because we are again dealing with pairs, and the same effect enters as in calculating the σ by use of pairs.

$$t = \frac{0.19 - 0}{0.082} \, 3^{\frac{1}{2}} = 4.01$$

For 2 df we get $t = 4.30$ from Table 9.3. Since the calculated value of t is less than the table value (for 0.99) we conclude that it is not proved that method X gives consistently higher results than method Y. However, additional data of similar nature, which increase df, may result in such a conclusion.

This latter procedure is quite useful in comparing a new technique with an old one on a group of different materials. Three differences produce two degrees of freedom, but usually in such a comparison, the longer series would be gradually accumulated.

SUMMARY. Significant differences of comparisons run in pairs may be quite useful even though averages of a large series are not available. We use the formula:

$$t = \frac{\bar{X} - E}{\sigma/n^{\frac{1}{2}}}$$

where E is the expected result. In pairs, E may be assumed as zero when we use $\bar{X}$ as the average difference of any set groups of paired values. The $n^{\frac{1}{2}}$ comes into the calculation because we are using paired values; n is the number of pairs.

9.5 DEGREES OF FREEDOM (df)

We have used the term, degrees of freedom, and seen it listed in Table 9.3. We shall now explain it further.

In determining error, it is obvious that one value does not reveal anything. When we get two values (X_1, X_2) we begin to get a conception of error. For n values, we get several estimates of error. Twenty values (X) can all be different; they exhibit individually a degree of freedom. When we strike an average, $\bar{X}$, we have used a degree of freedom. Thus, n values of X give $(n - 1)$ degrees of freedom for estimating error. This concept is hard to define, but easy to illustrate.

Simple Groups

2 values X_1, X_2	give $(2 - 1)$ df
3 values X_1, X_2, X_3	give $(3 - 1)$ df
10 values	give 9 df
n values	give $(n - 1)$ df

Groups of Data: Examples

Case 1

5 values give 4 df $(n - 1)$ 6 values give 5 df $(m - 1)$

The system $5 + 6$ gives 9 df. Thus we derive the formula:

$$df = (n - 1) + (m - 1) = (n + m - 2) = 9$$

Case 2

2 values	1 df	6 values	5 df
8 values	7 df	n values	$(n - 1)$ df

The system $df = 1 + 7 + 5 + (n - 1)$.

Case 3

10 values   9 df   10 values   9 df

Differences of 10 pairs of values $= 9$ df. This is the *df of the differences.*

1 pair gives 1 df   10 pairs give 10 df

Note: This concept is also used for calculation of the *F* test.

10

COMPARISON OF PRECISIONS OF TWO SETS OF MEASUREMENTS
The F Test

"We can easily represent things as we wish them to be."

10.1 VARIANCE

In Section 9 we have been concerned with determining whether the averages of two sets of data have significant differences. Another factor for comparing two sets of data is to determine whether the precisions are different. We have used sigma as a measure of precision. If the sample size is large enough, this is useful both for precision and comparison of two precisions. However, another slight variation is more sound from a statistical viewpoint. Its concept is simple.

For large groups:

$$(\text{sigma})^2 = \text{variance, or } \sigma'^2 = \sigma^2 = \text{variance}$$

For any group of data:

$$\text{variance} = \frac{\text{summations of } (d)^2}{(n-1)} = \frac{\text{sum of squares}}{\text{df}}$$

These formulas comprise a definition of the term "variance," where d = deviations from arithmetic average, and n = number of measurements. Thus the sigma value is calculated on the basis of $n - 1$; *the squaring of sigma gives the variance.*

Or, *the variance* is calculated by taking the deviations from the average of the data, squaring those deviations, and dividing the sum of the squares by *one less* than the number of measurements in the set.

133

Variance is defined as the square of the standard deviation.

The numerator term in the calculation of sigma, or variance, is known as the *sum of squares* (SS). This means the summation of the squares of the differences between, say, average and individual values, and the summation includes all such calculations for all individuals, or groups — if we are working with groups.

The SS has one very useful characteristic — it is additive for any analysis of a large group of data. Each important portion may be individually calculated, including the total.

By this SS characteristic, we divide a group of data into various components, each of which represents the effects of a variable, or combination of variables or unaccounted for segments. Thus the analysis of variance is a useful but sometimes tedious tool.

In simple analysis of variance cases, the procedure is not complex. The giving of simple instructions is difficult. Take the case of a group of 10 values (X) for three specific instances, such as given in Table 10.1. We calculate the SS (sum of squares) of columns as follows:

$$\text{SS}_{(\text{Instance A})} = (\bar{X}_A - X_{A1})^2 + (\bar{X}_A - X_{A2})^2 + \cdots + (\bar{X}_A - X_{A10})^2$$

The arithmetic shortcut (avoiding taking differences) used earlier in this book gives

$$\text{SS}_{(\text{Instance A})} = X_{A1}^2 + X_{A2}^2 + \cdots + X_{A10}^2$$
$$- \left[\frac{(X_{A1} + X_{A2} + \cdots + X_{A10})^2}{n} \right]$$

There are three such calculations for our illustration. We get SS_A; SS_B; SS_C (columns). We get the SS for each instance or factor. Now we need the SS for all the data as a group. This gives us the total SS for which we must account in our overall analysis of variance. When $n = 10$ values per column and $N = 3(10) = 30$ values for all data, the equation for all data entries is:

$$\text{total SS} = (X_{A1}^2 + X_{A2}^2 + \cdots + X_{C10}^2) - \left[\frac{(X_{A1} + X_{A2} + \cdots + X_{C10})^2}{N} \right]$$

This is the same equation as above for each column, but now expressed for all three columns combined. The general formula is:

$$\text{total SS} = \sum(X^2) - \frac{(\sum X)^2}{N}$$

where X stands for *each* individual entry. Our (analysis of variance) separate SS values must total to the above total SS.

Table 10.1 Data for Simple Analysis of Variance

Determination	Instance A	Instance B	Instance C
1	X_{A1}	X_{B1}	X_{C1}
2	X_{A2}	X_{B2}	X_{C2}
3	X_{A3}	X_{B3}	X_{C3}
4	X_{A4}	X_{B4}	X_{C4}
5	X_{A5}	X_{B5}	X_{C5}
6	X_{A6}	X_{B6}	X_{C6}
7	X_{A7}	X_{B7}	X_{C7}
8	X_{A8}	X_{B8}	X_{C8}
9	X_{A9}	X_{B9}	X_{C9}
10	X_{A10}	X_{B10}	X_{C10}
$n = 10$			
Total	ΣX_A	ΣX_B	ΣX_C
Average $\bar{X} = \dfrac{\text{Total}}{10}$	$\bar{X}_A$	$\bar{X}_B$	$\bar{X}_C$

We now calculate the SS for comparison of each level (between-levels) for A, B, C columns:

$$SS_{(\text{all levels})} = \left[\frac{(\sum X_A)^2}{n} + \frac{(\sum X_B)^2}{n} + \frac{(\sum X_C)^2}{n}\right] - \frac{(\sum X)^2}{N}$$

This gives us the SS for the three columns or a measure of the importance of levels A, B, and C represented in the columns.

If our "line" data were another variable (rather than the replication here used) we would repeat the SS for comparison of each level of the line variable. This would give the importance of levels expressed on the lines of the tabulation.

The inherent variation of the whole experiment (perhaps the experimental error) must next be estimated. It may be calculated by difference of the total SS minus the line SS and column SS as above calculated. Let us illustrate these steps for calculation of sum of squares by actual data.

DATA FROM OUR EXAMPLE E10-6 (SECTION 10.5). We have three shifts of operation (A, B, C), each with male (M) and female (F) crews. Our table entries are the productive output quality evaluations for 2 weeks, basis 100 expected, but coded by subtracting 100. (For discussion of coding, you

will await Section 10.4.) That is, an entry of 4 means a quality of 104, or 4%
to the good. See Table 10.2. According to this table,

average A shifts decoded = 103.6
average B shifts decoded = 104.0
average C shifts decoded = 102.4
average male crews decoded = 103.3
average female crews decoded = 102.3

and

$$\sum X = 45 + 27 + 30 + 20 + 25 + 23 = 170$$

$$\frac{\text{between shifts SS}}{\text{summating F \& M}} = \left[\frac{(45 + 27)^2}{20} + \frac{(30 + 20)^2}{20} + \frac{(25 + 23)^2}{20}\right] - \frac{170^2}{60}$$

$$SS = 499.4 - 481.7 = 17.7$$

$$df = 3 \text{ shifts} = 3 - 1 = 2$$

$$\frac{\text{between M \& F SS}}{\text{summating shifts}} = \left[\frac{(45 + 30 + 25)^2}{30} + \frac{(27 + 20 + 23)^2}{30}\right] - \frac{170^2}{60}$$
$$1 + 2 + 3$$
$$SS = 496.7 - 481.7 = 15.0$$

$$df(M \& F) = 2 - 1 = 1$$

$$\text{Total SS} = [(4^2 + 4^2 + \cdots + 2^2 + 3^2 + 3^2)] - \frac{170^2}{60} \qquad df = 59$$

$$= 652 - 481.7 = 170.3$$

$$\text{Residual SS (by difference)} = 170.3 - 17.7 - 15.0 = 137.6$$

$$df = 59 - 2 - 1 = 56$$

The problem here is that a large proportion of our SS rests in the
"residual" or normally the "error"; but it also includes the SS "between
days." We must look for the cause, as we shall later in Section 10.5, of the
same Example E10-6, where we account for only a little more of it. The
above calculations of SS show the simplicity of the calculation — our
purpose of the moment.

10.2   THE VARIANCE RATIO (THE *F* TEST)

By taking the variances of two sets of data and striking a ratio, we can determine whether there is a significant difference in the precisions of the two
sets of data.

Table 10.2   Quality Evaluation Related to Shift, Date and Operator

	Shift A		Shift B		Shift C	
Day	Female	Male	Female	Male	Female	Male
M	4	4	2	2	4	4
Tu	4	4	2	3	3	4
W	5	2	4	2	1	0
Th	5	2	6	2	2	2
F	3	2	8	5	2	3
M	6	3	3	0	2	0
Tu	4	2	3	0	3	2
W	4	5	0	4	3	2
Th	4	1	1	2	4	3
F	6	2	1	0	1	3
Σ columns (ΣX)	45	27	30	20	25	23
n	10	10	10	10	10	10
N	60	60	60	60	60	60
$\Sigma(X)^2$	211	87	144	66	73	71

This ratio, referred to as F, must be compared to theoretical values (Table 10.3) to determine whether two variances differ. The critical values in this table depend upon their parameters:

1. The desired confidence, G.
2. The degrees of freedom (df) for the variance in the numerator (m).
3. The degrees of freedom (df) for the variance in the denominator (n).

For the cases covered herein, df is ($n - 1$) where n is the number of data points used to estimate the corresponding variance.

Since this is a ratio, there may be some question as to which variance to place in the numerator. The rules follow.

Case 1: If the experiment is such that one of the variances is expected to be larger, not smaller, then this value must be placed in the numerator. If the modification of the experiment or comparison is supposed to improve precision, then the variance belonging to the original procedure must go in the numerator, even if numerically this variance is the smaller of the two. The corresponding critical value can be read directly from Table 10.3 and the expected confidence is then G from the table.

Case 2: Otherwise, put the larger variance in the numerator. In this case the appropriate confidence must be determined from the value ($2G - 1.0$).

For example, if $m = 1$, and $n = 9$, the critical value 3.36 opposite $G = 90$ is the critical value for 0.8 confidence ($2G - 1 = 2(0.90) - 1.0 = 0.80$).

Cases 1 and 2 above really refer to a one-tailed comparison (Case 1); or a two-tailed comparison (Case 2).

If the F *value exceeds that from Table 10.3, then there is evidence of a difference in precision of the two sets of data; or there is evidence that the presumed hypothesis is wrong.* A study of the F value table shows that unless one of the observed standard deviations is about 50% greater than the other, moderately sized sets will not provide convincing evidence of a difference in precision. To improve this discriminatory ability, larger sample sizes are required. Let us proceed to examples.

EXAMPLE E10-1.   In Figure 8.3 we have probability plots of three variables, A, B, and C, in lots of tubing. It is possible to read off sigma for these lots from these curves (the difference between 84% and 50%) as follows:

   Lot A, $\sigma = 0.007$ Lot B, $\sigma = 0.005$ Lot C, $\sigma = 0.0019$

These data represent a large sampling (n is large, perhaps over 100 for each lot).

Formulas for the Use of the F Table 10.3

In the F table (Table 10.3), G is the probability that a variance ratio with m degrees of freedom in the numerator and n in the denominator will not exceed the entry in the body of the table.

The F table gives values of the ratio of two variances for the degrees of freedom of them. If the calculated F value is greater than the table value, then there is evidence of a *difference in precision* of the two sets of data. The confidence of the decision is available from the F value table, as follows:

   0.995 value exceeded, very positive confidence
   0.95 value exceeded, confidence is acceptable
   0.90 value exceeded, confidence has some question

$$\text{variance} = \frac{\text{summation of } (d)^2}{(n - 1)}$$

$$F = \frac{\text{variance}_x}{\text{variance}_y} = \frac{\sigma_x^2}{\sigma_y^2}$$

In the table, m is the df of the numerator, and n is the df of the denominator. See Section 10.2 for further instructions.

Table 10.3 The Variance Ratio F (From E. B. Wilson, *An Introduction to Scientific Research*, McGraw-Hill, 1952, by permission of McGraw-Hill Book Company. Copyright 1952 by McGraw-Hill Book Company.)

									m										
G	n	1	2	3	4	5	6	7	8	9	10	12	15	20	30	60	120	∞
0.90	1	39.9	49.5	53.6	55.8	57.2	58.2	58.9	59.4	59.9	60.2	60.7	61.2	61.7	62.3	62.8	63.1	63.3
0.95		161	200	216	225	230	234	237	239	241	242	244	246	248	250	252	253	254
0.975		648	800	864	900	922	937	948	957	963	969	977	985	993	1000	1010	1010	1020
0.99		4,050	5,000	5,400	5,620	5,760	5,860	5,930	5,980	6,020	6,060	6,110	6,160	6,210	6,260	6,310	6,340	6,370
0.995		16,200	20,000	21,600	22,500	23,100	23,400	23,700	23,900	24,100	24,200	24,400	24,600	24,800	25,000	25,200	25,400	25,500
0.90	2	8.53	9.00	9.16	9.24	9.29	9.33	9.35	9.37	9.38	9.39	9.41	9.42	9.44	9.46	9.47	9.48	9.49
0.95		18.5	19.0	19.2	19.2	19.3	19.3	19.4	19.4	19.4	19.4	19.4	19.4	19.5	19.5	19.5	19.5	19.5
0.975		38.5	39.0	39.2	39.2	39.3	39.3	39.4	39.4	39.4	39.4	39.4	39.4	39.4	39.5	39.5	39.5	39.5
0.99		98.5	99.0	99.2	99.2	99.3	99.3	99.4	99.4	99.4	99.4	99.4	99.4	99.4	99.5	99.5	99.5	99.5
0.995		199	199	199	199	199	199	199	199	199	199	199	199	199	199	199	199	199
0.90	3	5.54	5.46	5.39	5.34	5.31	5.28	5.27	5.25	5.24	5.23	5.22	5.20	5.18	5.17	5.15	5.14	5.13
0.95		10.1	9.55	9.28	9.12	9.01	8.94	8.89	8.85	8.81	8.79	8.74	8.70	8.66	8.62	8.57	8.55	8.53
0.975		17.4	16.0	15.4	15.1	14.9	14.7	14.6	14.5	14.5	14.4	14.3	14.3	14.2	14.1	14.0	13.9	13.9
0.99		34.1	30.8	29.5	28.7	28.2	27.9	27.7	27.5	27.3	27.2	27.1	26.9	26.7	26.5	26.3	26.2	26.1
0.995		55.6	49.8	47.5	46.2	45.4	44.8	44.4	44.1	43.9	43.7	43.4	43.1	42.8	42.5	42.1	42.0	41.8
0.90	4	4.54	4.32	4.19	4.11	4.05	4.01	3.98	3.95	3.93	3.92	3.90	3.87	3.84	3.82	3.79	3.78	3.76
0.95		7.71	6.94	6.59	6.39	6.26	6.16	6.09	6.04	6.00	5.96	5.91	5.86	5.80	5.75	5.69	5.66	5.63
0.975		12.2	10.6	9.98	9.60	9.36	9.20	9.07	8.98	8.90	8.84	8.75	8.66	8.56	8.46	8.36	8.31	8.26
0.99		21.2	18.0	16.7	16.0	15.5	15.2	15.0	14.8	14.7	14.5	14.4	14.2	14.0	13.8	13.7	13.6	13.5
0.995		31.3	26.3	24.3	23.2	22.5	22.0	21.6	21.4	21.1	21.0	20.7	20.4	20.2	19.9	19.6	19.5	19.3
0.90	5	4.06	3.78	3.62	3.52	3.45	3.40	3.37	3.34	3.32	3.30	3.27	3.24	3.21	3.17	3.14	3.12	3.11
0.95		6.61	5.79	5.41	5.19	5.05	4.95	4.88	4.82	4.77	4.74	4.68	4.62	4.56	4.50	4.43	4.40	4.37
0.975		10.0	8.43	7.76	7.39	7.15	6.98	6.85	6.76	6.68	6.62	6.52	6.43	6.33	6.23	6.12	6.07	6.02
0.99		16.3	13.3	12.1	11.4	11.0	10.7	10.5	10.3	10.2	10.1	9.89	9.72	9.55	9.38	9.20	9.11	9.02
0.995		22.8	18.3	16.5	15.6	14.9	14.5	14.2	14.0	13.8	13.6	13.4	13.1	12.9	12.7	12.4	12.3	12.1

139

df = 6

prob																	
0.90	2.72	2.74	2.76	2.80	2.84	2.87	2.90	2.94	2.96	2.98	3.01	3.05	3.11	3.18	3.29	3.46	3.78
0.95	3.67	3.70	3.74	3.81	3.87	3.94	4.00	4.06	4.10	4.15	4.21	4.28	4.39	4.53	4.76	5.14	5.99
0.975	4.85	4.90	4.96	5.07	5.17	5.27	5.37	5.46	5.52	5.60	5.70	5.82	5.99	6.23	6.60	7.26	8.81
0.99	6.88	6.97	7.06	7.23	7.40	7.56	7.72	7.87	7.98	8.10	8.26	8.47	8.75	9.15	9.78	10.9	13.7
0.995	8.88	9.00	9.12	9.36	9.59	9.81	10.0	10.2	10.4	10.6	10.8	11.1	11.5	12.0	12.9	14.5	18.6

df = 7

prob																	
0.90	2.47	2.49	2.51	2.56	2.59	2.63	2.67	2.70	2.72	2.75	2.78	2.83	2.88	2.96	3.07	3.26	3.59
0.95	3.23	3.27	3.30	3.38	3.44	3.51	3.57	3.64	3.68	3.73	3.79	3.87	3.97	4.12	4.35	4.74	5.59
0.975	4.14	4.20	4.25	4.36	4.47	4.57	4.67	4.76	4.82	4.90	4.99	5.12	5.29	5.52	5.89	6.54	8.07
0.99	5.65	5.74	5.82	5.99	6.16	6.31	6.47	6.62	6.72	6.84	6.99	7.19	7.46	7.85	8.45	9.55	12.2
0.995	7.08	7.19	7.31	7.53	7.75	7.97	8.18	8.38	8.51	8.68	8.89	9.16	9.52	10.1	10.9	12.4	16.2

df = 8

prob																	
0.90	2.29	2.31	2.34	2.38	2.42	2.46	2.50	2.54	2.56	2.59	2.62	2.67	2.73	2.81	2.92	3.11	3.46
0.95	2.93	2.97	3.01	3.08	3.15	3.22	3.28	3.35	3.39	3.44	3.50	3.58	3.69	3.84	4.07	4.46	5.32
0.975	3.67	3.73	3.78	3.89	4.00	4.10	4.20	4.30	4.36	4.43	4.53	4.65	4.82	5.05	5.42	6.06	7.57
0.99	4.86	4.95	5.03	5.20	5.36	5.52	5.67	5.81	5.91	6.03	6.18	6.37	6.63	7.01	7.59	8.65	11.3
0.995	5.95	6.06	6.18	6.40	6.61	6.81	7.01	7.21	7.34	7.50	7.69	7.95	8.30	8.81	9.60	11.0	14.7

df = 9

prob																	
0.90	2.16	2.18	2.21	2.25	2.30	2.34	2.38	2.42	2.44	2.47	2.51	2.55	2.61	2.69	2.81	3.01	3.36
0.95	2.71	2.75	2.79	2.86	2.94	3.01	3.07	3.14	3.18	3.23	3.29	3.37	3.48	3.63	3.86	4.26	5.12
0.975	3.33	3.39	3.45	3.56	3.67	3.77	3.87	3.96	4.03	4.10	4.20	4.32	4.48	4.72	5.08	5.71	7.21
0.99	4.31	4.40	4.48	4.65	4.81	4.96	5.11	5.26	5.35	5.47	5.61	5.80	6.06	6.42	6.99	8.02	10.6
0.995	5.19	5.30	5.41	5.62	5.83	6.03	6.23	6.42	6.54	6.69	6.88	7.13	7.47	7.96	8.72	10.1	13.6

df = 10

prob																	
0.90	2.06	2.08	2.11	2.15	2.20	2.24	2.28	2.32	2.35	2.38	2.41	2.46	2.52	2.61	2.73	2.92	3.29
0.95	2.54	2.58	2.62	2.70	2.77	2.84	2.91	2.98	3.02	3.07	3.14	3.22	3.33	3.48	3.71	4.10	4.96
0.975	3.08	3.14	3.20	3.31	3.42	3.52	3.62	3.72	3.78	3.85	3.95	4.07	4.24	4.47	4.83	5.46	6.94
0.99	3.91	4.00	4.08	4.25	4.41	4.56	4.71	4.85	4.94	5.06	5.20	5.39	5.64	5.99	6.55	7.56	10.0
0.995	4.64	4.75	4.86	5.07	5.27	5.47	5.66	5.85	5.97	6.12	6.30	6.54	6.87	7.34	8.08	9.43	12.8

df = 12

prob																	
0.90	1.90	1.93	1.96	2.01	2.06	2.10	2.15	2.19	2.21	2.24	2.28	2.33	2.39	2.48	2.61	2.81	3.18
0.95	2.30	2.34	2.38	2.47	2.54	2.62	2.69	2.75	2.80	2.85	2.91	3.00	3.11	3.26	3.49	3.89	4.75
0.975	2.72	2.79	2.85	2.96	3.07	3.18	3.28	3.37	3.44	3.51	3.61	3.73	3.89	4.12	4.47	5.10	6.55
0.99	3.36	3.45	3.54	3.70	3.86	4.01	4.16	4.30	4.39	4.50	4.64	4.82	5.06	5.41	5.95	6.93	9.33
0.995	3.90	4.01	4.12	4.33	4.53	4.72	4.91	5.09	5.20	5.35	5.52	5.76	6.07	6.52	7.23	8.51	11.8

df	p																	
15	0.90	1.76	1.79	1.82	1.87	1.92	1.97	2.02	2.06	2.09	2.12	2.16	2.21	2.27	2.36	2.49	2.70	3.07
	0.95	2.07	2.11	2.16	2.25	2.33	2.40	2.48	2.54	2.59	2.64	2.71	2.79	2.90	3.06	3.29	3.68	4.54
	0.975	2.40	2.46	2.52	2.64	2.76	2.86	2.96	3.06	3.12	3.20	3.29	3.41	3.58	3.80	4.15	4.77	6.20
	0.99	2.87	2.96	3.05	3.21	3.37	3.52	3.67	3.80	3.89	4.00	4.14	4.32	4.56	4.89	5.42	6.36	8.68
	0.995	3.26	3.37	3.48	3.69	3.88	4.07	4.25	4.42	4.54	4.67	4.85	5.07	5.37	5.80	6.48	7.70	10.8
20	0.90	1.61	1.64	1.68	1.74	1.79	1.84	1.89	1.94	1.96	2.00	2.04	2.09	2.16	2.25	2.38	2.59	2.97
	0.95	1.84	1.90	1.95	2.04	2.12	2.20	2.28	2.35	2.39	2.45	2.51	2.60	2.71	2.87	3.10	3.49	4.35
	0.975	2.09	2.16	2.22	2.35	2.46	2.57	2.68	2.77	2.84	2.91	3.01	3.13	3.29	3.51	3.86	4.46	5.87
	0.99	2.42	2.52	2.61	2.78	2.94	3.09	3.23	3.37	3.46	3.56	3.70	3.87	4.10	4.43	4.94	5.85	8.10
	0.995	2.69	2.81	2.92	3.12	3.32	3.50	3.68	3.85	3.96	4.09	4.26	4.47	4.76	5.17	5.82	6.99	9.94
30	0.90	1.46	1.50	1.54	1.61	1.67	1.72	1.77	1.82	1.85	1.88	1.93	1.98	2.05	2.14	2.28	2.49	2.88
	0.95	1.62	1.68	1.74	1.84	1.93	2.01	2.09	2.16	2.21	2.27	2.33	2.42	2.53	2.69	2.92	3.32	4.17
	0.975	1.79	1.87	1.94	2.07	2.20	2.31	2.41	2.51	2.57	2.65	2.75	2.87	3.03	3.25	3.59	4.18	5.57
	0.99	2.01	2.11	2.21	2.39	2.55	2.70	2.84	2.98	3.07	3.17	3.30	3.47	3.70	4.02	4.51	5.39	7.56
	0.995	2.18	2.30	2.42	2.63	2.82	3.01	3.18	3.34	3.45	3.58	3.74	3.95	4.23	4.62	5.24	6.35	9.18
60	0.90	1.29	1.35	1.40	1.48	1.54	1.60	1.66	1.71	1.74	1.77	1.82	1.87	1.95	2.04	2.18	2.39	2.79
	0.95	1.39	1.47	1.53	1.65	1.75	1.84	1.92	1.99	2.04	2.10	2.17	2.25	2.37	2.53	2.76	3.15	4.00
	0.975	1.48	1.58	1.67	1.82	1.94	2.06	2.17	2.27	2.33	2.41	2.51	2.63	2.79	3.01	3.34	3.93	5.29
	0.99	1.60	1.73	1.84	2.03	2.20	2.35	2.50	2.63	2.72	2.82	2.95	3.12	3.34	3.65	4.13	4.98	7.08
	0.995	1.69	1.83	1.96	2.19	2.39	2.57	2.74	2.90	3.01	3.13	3.29	3.49	3.76	4.14	4.73	5.80	8.49
120	0.90	1.19	1.26	1.32	1.41	1.48	1.54	1.60	1.65	1.68	1.72	1.77	1.82	1.90	1.99	2.13	2.35	2.75
	0.95	1.25	1.35	1.43	1.55	1.66	1.75	1.83	1.91	1.96	2.02	2.09	2.18	2.29	2.45	2.68	3.07	3.92
	0.975	1.31	1.43	1.53	1.69	1.82	1.94	2.05	2.16	2.22	2.30	2.39	2.52	2.67	2.89	3.23	3.80	5.15
	0.99	1.38	1.53	1.66	1.86	2.03	2.19	2.34	2.47	2.56	2.66	2.79	2.96	3.17	3.48	3.95	4.79	6.85
	0.995	1.43	1.61	1.75	1.98	2.19	2.37	2.54	2.71	2.81	2.93	3.09	3.28	3.55	3.92	4.50	5.54	8.18
∞	0.90	1.00	1.17	1.24	1.34	1.42	1.49	1.55	1.60	1.63	1.67	1.72	1.77	1.85	1.94	2.08	2.30	2.71
	0.95	1.00	1.22	1.32	1.46	1.57	1.67	1.75	1.83	1.88	1.94	2.01	2.10	2.21	2.37	2.60	3.00	3.84
	0.975	1.00	1.27	1.39	1.57	1.71	1.83	1.94	2.05	2.11	2.19	2.29	2.41	2.57	2.79	3.12	3.69	5.02
	0.99	1.00	1.32	1.47	1.70	1.88	2.04	2.18	2.32	2.41	2.51	2.64	2.80	3.02	3.32	3.78	4.61	6.63
	0.995	1.00	1.36	1.53	1.79	2.00	2.19	2.36	2.52	2.62	2.74	2.90	3.09	3.35	3.72	4.28	5.30	7.88

I have reason to believe that the process changes from A to B and from B to C should greatly improve the spread of results. So I have Case 1 and the G value becomes my confidence value, as follows:

comparing lots A and B variance $A = 0.007^2$ variance $B = 0.005^2$

$$\text{ratio } F_{A+B} = \frac{0.007^2}{0.005^2} = \frac{49}{25} = 2.0$$

From Table 10.3 for 120 df in both numerator and denominator, F_{A+B} is

$G = 0.90$	Confidence $= 0.90$	$F = 1.26$
$= 0.95$	$= 0.95$	$= 1.35$
$= 0.975$	$= 0.975$	$= 1.43$
$= 0.99$	$= 0.99$	$= 1.53$
$= 0.995$	$= 0.995$	$= 1.61$

The ratio F_{A+B} is measured as 2.0. It exceeds even the 0.995 confidence F table value. So my hypothesis that lot B is much more precise than lot A must stand.

My second hypothesis, that lot C is much more precise than lot B, calculates as follows when variance $B = 0.005^2$ and variance $C = 0.0019^2$

$$\text{ratio } F_{B+C} = \frac{0.005^2}{0.0019^2} = \frac{25}{3.6} = 7$$

This far exceeds the table values, and the hypothesis is supported.

EXAMPLE E10-2.   In the strength of glass rods example (Figure 9.2), we gave data for two lots, A and B. We now add data for three more lots, C, D, and E. These are plotted in Figure 10.1. Table 10.4 summarizes the data.

Just the slopes, and the fit of the data to the curves (Figures 9.2 and 10.1) show much difference in precision. However, let us look at the vari-

Table 10.4   Glass Rod Strengths

Test Lot	Average Strength	Sigma from Figures 9.2 and 10.1	Sample (n)
(1) Lot B	24.5	4.7	19
(2) Lot A	20.0	3.2	10
(3) Lot C	11.8	1.4	20
(4) Lot D	11.5	1.7	10
(5) Lot E	8.8	1.3	10

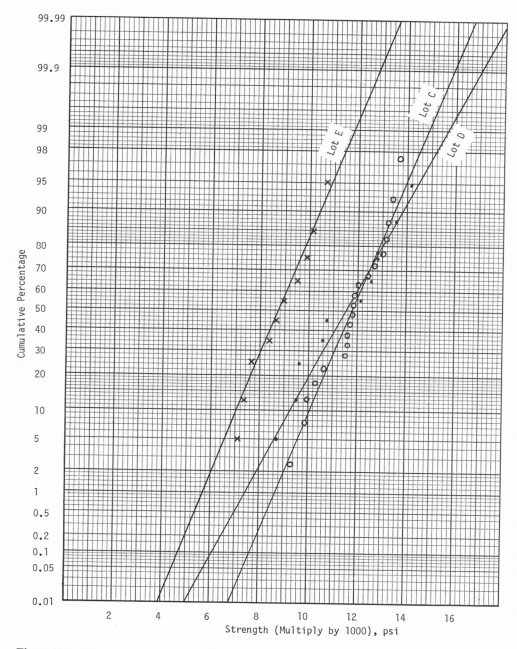

Figure 10.1 Transverse strength of glass rods, lots *C*, *D*, and *E*. See Example E10-2.

ance ratios. First, I must state the hypotheses we believe from the experimental work.

I expect that test lot 2 should average below lot 1, but see no reason to believe that the precision should be improved.

I expect that lot 3 should have much greater precision than lot 1.
I see no reason to expect lot 5 to have improved precision over lot 4.
I have no further speculations.

We must tabulate these comparisons.

$$\text{Lots 1 and 2} \qquad F = \frac{4.7^2}{3.2^2} = \frac{22.1}{10.2} = 2.17$$

$$\text{Lots 1 and 3} \qquad F = \frac{4.7^2}{1.4^2} = \frac{22.1}{1.96} = 11.3$$

$$\text{Lots 4 and 5} \qquad F = \frac{1.7^2}{1.3^2} = \frac{2.89}{1.69} = 1.71$$

$$\text{Lots 3 and 4} \qquad F = \frac{1.7^2}{1.4^2} = \frac{2.89}{1.96} = 1.48$$

$$\text{Lots 3 and 5} \qquad F = \frac{1.4^2}{1.3^2} = \frac{1.96}{1.69} = 1.16$$

Let us tabulate these values of F, versus the Table F values, and the appropriate confidence. See Table 10.5.

Table 10.5   The F Test on Strength of Glass Compositions

Lots	Calculated F Value	Table F (G = 0.95) Case 1	Table F (G = 0.975*) Case 2
1 and 2	2.17	—	3.7
1 and 3	11.3	2.2	—
4 and 5	1.71	—	4.0
3 and 4	1.48	—	2.9
3 and 5	1.16	—	3.7

*Confidence equals $2G - 1$ or 0.95.

On this basis, only the comparison of lots 1 and 3 shows significance in their differences of precision. If we made other pairs of comparisons, say lots 2 and 4 or lots 1 and 4, we would get significance shown.

One sees here that the F test is very critical. Many samples are required for a determination of the variance (thus the σ) of a set of data. More about the sample size later. (Section 10.6).

EXAMPLE E10-3. Another example will show the usage of the F test to compare two series. In the fourth example under Section 9.4, Table 9.7, we show the data, part of which we use again in Table 10.6. There are no hypotheses — we have no reason to lead us to believe that the variances of the data should be different. So for 0.95 confidence, we use $G = 0.975$; for 0.90 confidence, we use $G = 0.95$. We thus conclude that there is no evidence to suspect a difference in precision between the two workers.

Table 10.6 Use of F Test to Compare Two Series of Analyses

	Analyst	
	I	*II*
Specific results	59.09	59.51
	59.17	59.75
	59.27	59.61
	59.13	59.60
	59.10	
	59.14	
Average	59.15	59.62
Sums of squares (SS)	0.0214	0.0295
Degrees of freedom (df)	5	3
Variance (SS)/(df)	0.00428	0.00983
Ratio of variances, F	2.30	
Critical value:		
(5% level) $G = 0.975$	7.76	
(10% level) $G = 0.95$	5.41	

10.3 ANALYSIS OF VARIANCE

In investigating differences among several variables, each pair may be subjected to the t test. But this is a lot of work for several variables, each run at several levels. Further, it can be shown that this procedure would result in a high risk of one or more incorrect conclusions. The analysis of variance procedure may be useful but is detailed. We gain more precision in our statistical procedure by using all available data in our analysis.

The analysis of variance depends upon separation of the variance into parts (each due to some specific variable or source of error). Internal variation of single populations may be evaluated and compared to other single populations.

Manual calculations may be time consuming.

The *F* test is usable on a sum of squares of the differences basis. Variance was defined as:

$$\text{variance} = \text{sigma}^2$$

but

$$\text{sigma}^2 = \frac{\sum(\text{differences})^2}{n - 1}$$

where $\sum(\text{differences})^2 = $ sum of squares

$$F = \frac{\text{variance}}{\text{variance}}$$

$$F = \frac{\sum(\text{differences})^2/(n - 1)}{\sum(\text{differences})^2/(n - 1)}$$

Two cases exist. If $(n - 1)$ for all data groups is the same, then

$$F = \frac{\sum(\text{differences})^2}{\sum(\text{differences})^2}$$

If $(n - 1)$ differs, then the full formula above must be used. It is common to call the value *mean square*,

$$\text{mean square} = \frac{\sum(\text{differences})^2}{\text{df}} = \frac{\text{sum of squares}}{\text{df}}$$

$$F = \frac{\text{mean square}}{\text{mean square}}$$

It is also useful to say that the variance estimate based on means (averages) is called a *between-means variance*. The estimate based on all individual data is called the *within individuals*, or better, the *residual variance*.

We need not calculate sigma, as we have shown in Example E10-3; but we understand its meaning clearly and may find its calculation useful.

The sums of squares is additive, so we may take data, break out certain experimental facets, and then evaluate and rank the factors, variances, discrepancies, etc.

EXAMPLE E10-4.   A four by four Latin square for a glass melting experiment is shown in Table 10.7. The values listed in Table 10.8 are for glass quality (not replications) as judged by the count of very small blisters (seeds) for unit weight of samples from four slightly different compositions of glass (I, II, III, IV), from four different glass furnaces (1, 2, 3, 4), and produced at four different temperatures of melting (A, B, C, D). The data are hypothetical. Sums of squares are calculated for:

glass composition (I, II, III, IV)

$$\frac{(207)^2}{4} + \frac{(214)^2}{4} + \frac{(196)^2}{4} + \frac{(219)^2}{4} - \frac{(836)^2}{16} = 43755.5 - 43681 = 74.5$$

Table 10.7   Latin Square, 4 × 4, for a Glass Melting Experiment

	I	II	III	IV	Total
1	A	B	C	D	ABCD
2	D	A	B	C	ABCD
3	C	D	A	B	ABCD
4	B	C	D	A	ABCD
Total	ABCD	ABCD	ABCD	ABCD	4(ABCD)
Averages		Four A's			
Averages		Four B's			
Averages		Four C's			
Averages		Four D's			

Table 10.8   A Glass Melting Experiment — Original Data

Melting Furnace	Glass Composition				Total of Lines (Furnace Used)
	I	II	III	IV	
1	A 81	B 41	C 44	D 53	219
2	D 38	A 97	B 42	C 49	226
3	C 31	D 43	A 67	B 36	177
4	B 57	C 33	D 43	A 81	214
Total Columns (Composition)	207	214	196	219	836

Total temperature of operation

$A = 326$

$B = 176$

$C = 157$

$D = 177$

$ABCD = 836$

melting furnace (1, 2, 3, 4)

$$\frac{(219)^2}{4} + \frac{(226)^2}{4} + \frac{(177)^2}{4} + \frac{(214)^2}{4} - \frac{(836)^2}{16} = 44040.5 - 43681 = 359.5$$

temperature of operation (A, B, C, D)

$$\frac{(326)^2}{4} + \frac{(176)^2}{4} + \frac{(157)^2}{4} + \frac{(177)^2}{4} - \frac{(836)^2}{16} = 48307.5 - 43681 = 4626.5$$

total (all 16 data entries)

$$81^2 + 38^2 + \cdots + 36^2 + 81^2 - \frac{(836)^2}{16} = 49348 - 43681 = 5667$$

Now we must set up an analysis of variance (Table 10.9). In computing we will use as values df numerator = 3, df denominator = 6.

Confidence (figures from Case 1) = 0.90, 0.95, 0.975, 0.99, 0.995.
F ratios (from Table 10.3) = 3.29, 4.76, 6.60, 9.78, 12.9.

Note: SS are additive; MS are not additive.

Our F value for glass composition is 0.25. There are no significant differences in the groups.

For melting furnaces, our F value is 1.18; again no differences are significant.

But for melting temperatures, our F value is 15.3; this is a very great

Table 10.9 Analysis of Variance of a Glass Melting Experiment

Data Used	Source of Variations	Sum of Squares (SS)	df	Mean Square (MS)	F Ratio Related to Residual
All	All individuals	5667	15	—	—
Columns I, II, III, IV	Glass compositions	74.5	3	24.8	0.25
Lines 1, 2, 3, 4	Melting furnaces	359.5	3	119.8	1.18
A, B, C, D	Temperatures of operation	4626.5	3	1542.2	15.3
Residual	Any remaining	606.5	6	101.1	—
Total		5667	15	377.8	—

significance — a great difference exists between the chosen melting temperatures.

   Note: Our residual here is small, so we feel good.

   Suppose that we had approached this problem from the simpler concepts known before we talked of analysis of variance. We could have plotted all the 16 values on probability paper. This plot (Figure 10.2) shows some peculiar characteristics. We look for the reasons for the deviation from the linear, which are too consistent to be ignored. We go further.

   Suppose we approach the question of plotting the totals from the melting furnaces, the totals from the glass compositions, and the totals from the temperatures of operation on probability paper. We have only 4 such totals for each of the 3 above groupings. We must *calculate* averages and sigma. *Note:* We are calculating averages and sigmas for groups of 4.

   For glass composition, the totals are 196, 207, 214, 219 (from Table 10.8). The average is, of course, 209. Then

$$(209 - 196)^2 + (209 - 207)^2 + (209 - 214)^2 + (209 - 219)^2 = 298$$

$$\sigma = \left(\frac{298}{3}\right)^{\frac{1}{2}} = 10$$

   For temperatures of operation, the calculation is

$$(209 - 326)^2 + (209 - 176)^2 + (209 - 157)^2 + (209 - 177)^2 = 18506$$

$$\sigma = \left(\frac{18506}{3}\right)^{\frac{1}{2}} = 78.5$$

   For melting furnaces, we calculate

$$(209 - 177)^2 + (209 - 214)^2 + (209 - 219)^2 + (209 - 226)^2 = 1438$$

$$\sigma = \left(\frac{1438}{3}\right)^{\frac{1}{2}} = 22$$

   For all the data (grouped, not individual), the sum of squares of 3 above calculations is

$$298 + 18,506 + 1,438 = 20,242$$

$$\sigma = \left(\frac{20,242}{15}\right)^{\frac{1}{2}} = 36.7$$

   Now let us plot these 4 lines on probability paper. Figure 10.3 shows right away that the large sigma is from the temperatures of operation, and that the glass compositions are closely similar. The melting furnaces fall between these two cases.

   So a probability plot does prove useful — and visual.

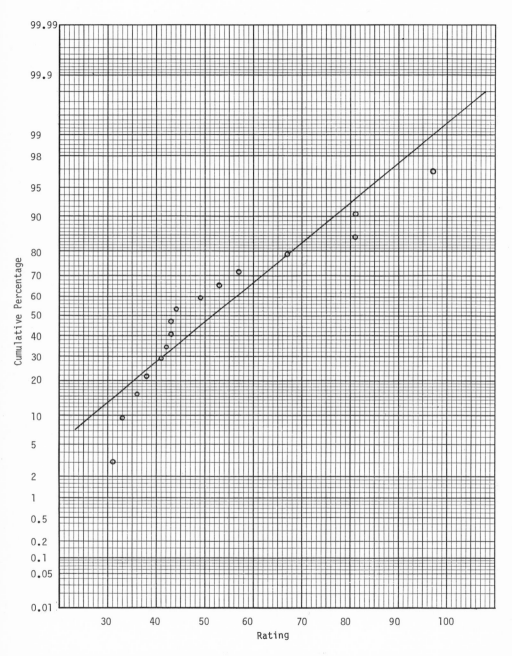

Figure 10.2 The individual data probability plot of a glass melting experiment. See Example E10-4.

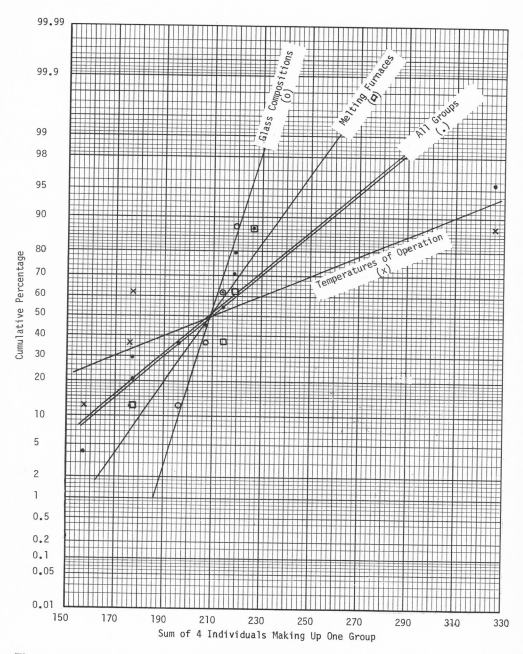

Figure 10.3 The sums of four individuals making up groups of a glass melting experiment. Lines were drawn after calculation of average and sigma. See text. Four points are not enough for a good line location in this case. See Example E10-4.

10.4 SHORT CUTS IN CALCULATIONS
FOR VARIANCE AND ANALYSIS OF VARIANCE

When we consider all the mathematical aspects of the calculation of variance, or the analysis of variance, we realize that the F test is a test of *ratios* of variances. Therefore, there are several arithmetic shortcuts (called coding) in the calculations:

  1. We may reduce each item by a constant amount (subtraction).
  2. We may increase each item by a constant amount (addition).
  3. We may multiply each item by a constant amount (multiplication).
  4. We may divide each item by a constant amount (division).
  5. We may combine any of the above.

  Since we are often squaring numbers, it is convenient to keep such numbers small, even if some come out negative (the squaring overcomes this problem).

  EXAMPLE E10-5. Take the case of the output of four workers, on three differently equipped production facilities, with each facility made available to each worker. The data for a complete factorial plan are found in Table 10.10. Calculated sigma for workers is 24.5; calculated sigma for facilities is 80.

  From the table itself we suspect that production facility 3 is far more efficient than is facility 1. We also feel that worker A is superior to worker C. The complete analysis of variance would help us in such evaluation. We want our table in smaller numbers for easy arithmetic (Table 10.11). We code by subtracting 50 units, and dividing by 10. Let X equal actual value, then coded value X' is

$$X' = \frac{X - 50}{10}$$

  Calculated sigma for workers here is 2.45; calculated sigma for facilities is 8.0. Note they are one-tenth those of Table 10.10 (our coding factor).

  Analysis of variance based on Table 10.11 will give the same result as that based on Table 10.10. Using values in Table 10.11, we get

Sum of squares — workers (columns):

$$\frac{6^2}{3} + \frac{3^2}{3} + \frac{0^2}{3} + \frac{3^2}{3} - \frac{(6 + 3 + 0 + 3)^2}{12} = 6$$

Sum of squares — facilities (lines):

$$\frac{-4^2}{4} + \frac{4^2}{4} + \frac{12^2}{4} - \frac{(4 + 4 + 12)^2}{12} = 32$$

Table 10.10 Production Output (Units) Versus Facility and Worker

Facility Number	Worker				Facility Total
	A	B	C	D	
1	30	70	30	30	160
2	80	50	40	70	240
3	100	60	80	80	320
Worker total	210	180	150	180	720

Table 10.11 Reduction by Coding of Table 10.10: Subtracting 50 Units From Each Value and Then Dividing the Result by 10

Facility Number	Worker				Facility Total
	A	B	C	D	
1	−2	2	−2	−2	−4
2	3	0	−1	2	4
3	5	1	3	3	12
Worker total	6	3	0	3	12

Sum of squares — total individuals:

$$(-2^2 + 3^2 + 5^2 + \cdots + 2^2 + 3^2)$$
$$- \frac{(-2 + 3 + 5 \cdots + 3)^2}{12} = (74) - (12) = 62$$

Note: You can do almost all of these calculations in your head or on a half sheet of paper.

The analysis of variance is given in Table 10.12. Our residual here is pretty large; we should not be too happy with it.

Based on the value of *F* ratio from Table 10.3 for $m = 2$ and $n = 6$ of 3.46 for 0.90 probability, and 5.14 for 0.95 probability, we cannot be sure of a difference in facilities.

Table 10.12 Analysis of Variance. Production Output Versus Facility and Worker (coded Data)

Source of Variance	Sum of Squares	df	Mean Square	F Ratio Related to Residual
Workers	6	3	2	0.50
Facilities	32	2	16	4.0
Residual*	24	6	4	
Total	62	11		

*Difference not accounted for by either workers or facilities.

We now see the conservative *F* test not proving what seems so obvious from the original data in Table 10.10.

Concerning coding of data for easier calculation, we have the following rules for *converting* the calculated variance or sigma *back to the original uncoded units*.

1. A standard deviation computed on coded data is affected by multiplication or division only — not by adding or subtracting a constant to data.

2. If coding involves multiplication or division, the inverse operation must be applied to the computed standard deviation to bring it back to the original units.

3. A variance computed on coded data must be multiplied by the square of the coding factor if division has been used in coding; or divided by the square of the coding factor, if multiplication was used.

In our above case, the coding factor was division by 10. Thus, to get standard deviation back to original units, we must multiply by 10.

10.5 INTERACTION STUDIED BY ANALYSIS OF VARIANCE

Interaction is a measure of the differential response caused by different levels of one factor at various levels of other factors. Crude plotting of results as a function of levels of other factors is always desirable. Roughly a crisscrossing of such plots warns one to be on the lookout for interaction. Rough parallelism gives some assurance of the lack of interaction. A commonplace case is the combined effect of time and temperature on reactions. Oftentimes, the same result is attainable by using a longer time and a lower temperature, or vice versa. Interacting factors might often invalidate the expected results from some combinations of such factors.

Experiments run at three or more levels are more likely to be useful in studying interaction. Two points serve only to locate a straight line. Three or more points may reveal a curve.

EXAMPLE E10-6. Entries in Table 10.13 are quality of output, basis expected 100, but coded by subtracting 100 (See Section 10.4 on coding) for two weeks, three shifts (A, B, C), with male (M) and female (F) crews of workers.

Table 10.13 Original Data for Output Related to Shift and Male-Female Crews

Day	Shift A Female	Shift A Male	Shift B Female	Shift B Male	Shift C Female	Shift C Male
M	4	4	2	2	4	4
Tu	4	4	2	3	3	4
W	5	2	4	2	1	0
Th	5	2	6	2	2	2
F	3	2	8	5	2	3
M	6	3	3	0	2	0
Tu	4	2	3	0	3	2
W	4	5	0	4	3	2
Th	4	1	1	2	4	3
F	6	2	1	0	1	3
Σ columns (ΣX)	45	27	30	20	25	23
n	10	10	10	10	10	10
N	60	60	60	60	60	60
$\Sigma(X)^2$	211	87	144	66	73	71

The average for shifts decoded is as follows: shift A = 103.6, shift B = 104.0, shift C = 102.4. The average for male workers decoded is 103.3 The average for female workers decoded = 102.3. Thus we have

$$\sum X = 45 + 27 + 30 + 20 + 25 + 23 = 170$$

$$\text{total SS} = [(4^2 + 4^2 + \cdots + 2^2 + 3^2 + 3^2)] - \frac{170^2}{60} \qquad df = 59$$

$$= 652 - 481.7 = 170.3$$

Enter results on line 1 of table of analysis of variance, (Table 10.14).

$$\frac{\text{between shifts SS}}{\text{summating F \& M}} = \left[\frac{(45 + 27)^2}{20} + \frac{(30 + 20)^2}{20} + \frac{(25 + 23)^2}{20}\right] - \frac{170^2}{60}$$

$$SS = 499.4 - 481.7 = 17.7$$

$$df = 3 \text{ shifts} = 3 - 1 = 2$$

Enter results on line 2 of Table 10.14.

$$\frac{\text{between M \& F SS}}{\substack{\text{summating shifts} \\ 1 + 2 + 3}} = \left[\frac{(45 + 30 + 25)^2}{30} + \frac{(27 + 20 + 23)^2}{30}\right] - \frac{170^2}{60}$$

$$SS = 496.7 - 481.7 = 15.0$$

$$df(M \& F) = 2 - 1 = 1$$

Enter results on line 3 of Table 10.14.

INTERACTION SS.

$$SS = \left[\frac{45^2}{10} + \frac{27^2}{10} + \frac{30^2}{10} + \frac{20^2}{10} + \frac{25^2}{10} + \frac{23^2}{10}\right] - \frac{170^2}{60} - 17.7 - 15.0$$

The 17.7 and 15.0 subtracted here have been corrected for SS shifts and crews which are included in the basic SS interaction calculation. Continuing, we have

$$SS = 520.8 - 481.7 - 17.7 - 15.0$$
$$SS = 520.8 - 514.4 = 6.4$$
$$df = 2 \times 1 = 2$$

Enter results on line 4 of Table 10.14.

RESIDUAL, BY DIFFERENCE.

$$SS = 170.3 - 17.7 - 15.0 - 6.4$$
$$SS = 131.2$$
$$df = 59 - 2 - 1 - 2 = 54$$

Enter results on line 5 of Table 10.14.

TOTAL SS.   Reenter on line 6 the figures on line 1 of Table 10.14.

As defined earlier, we calculate SS/df and call it mean square (MS).
The ratio F is MS/MS which is compared to the table F value for significances (with proper df values) as we have done in Section 10.2.

Our interaction of shifts and male-female crews is negative. There is no significance (n.s.).

Table 10.14 Analysis of Variance for Output with Interaction

Source of Variation	Sum of Squares	df	Mean Square	F Ratio
Total	170.3	59	—	—
Between shifts	17.7	2	8.85	3.64
Between male and female	15.0	1	15.0	6.17
Interaction (shifts $\times$ MF)	6.4	2	3.2	1.32 (n.s.)
Residual	131.2	54	2.43	
Total	170.3	59		

The designation (n.s.) means not significant.

The male-female crews figure, with F ratio of 6.17, is very significant compared with the table value, as follows:

$$\frac{df}{df} = \frac{1}{60}$$

confidence $= 0.95$ F ratio from Table 10.3 $= 4.00$
            $= 0.975$ $= 5.29$
            $= 0.99$ $= 7.08$

There is a difference between M and F crews. The between-shifts F ratio is 3.64. We compare to table values

$$\frac{df}{df} = \frac{2}{60}$$

confidence $= 0.95$ F ratio from Table 10.3 $= 3.15$
            $= 0.975$ $= 3.93$

There is a significant difference for the shifts.

EXAMPLE E10-7. Analysis of variance with interaction — 4 variables, 3 levels, 3 replications.

Our illustration is a problem of 4 variables, each studied at 3 levels, with 3 replications. This gives a total of 3^5 experimental results — or 243 entries in a table. The original data might well be shown in the form of Table 10.15, which is only partially filled in because the original data are too cumbersome to enter here since they are not necessary for our purpose. Each asterisk represents an entry in Table 10.15. Thus there are 3 entries per square.

Table 10.15 The Original Data Layout for Interaction; 4 Variables, 3 Levels Each, 3 Replications

		A_1			A_2			A_3		
		B_1	B_2	B_3	B_1	B_2	B_3	B_1	B_2	B_3
	C_1	***								
D_1	C_2	***								
	C_3	***								
	C_1	***								
D_2	C_2									
	C_3									***
	C_1									***
D_3	C_2									***
	C_3									***

When $n = 3$ and $N = 243$, the degrees of freedom (df) are as follows:

The full system $243 - 1 = 242$

For *A* calculation $3 - 1 = 2$ For *C* calculation $3 - 1 = 2$
For *B* calculation $3 - 1 = 2$ For *D* calculation $3 - 1 = 2$

Any two of above: AB, etc $= 2 \times 2 = 4$
Any three of above: ABC, etc $= 2 \times 2 \times 2 = 8$

Without our data, *we will assume an SS of 1725 for this table.*

Various subtables, each containing a summation of *all* the data in Table 10.15, can be built up. These may take the form of eliminating certain variables by taking their sums at levels 1, 2, and 3 and dropping their identity so as to study the remaining variables. When summations are made, the calculations using the summated data are *always* related to the original number of data entries — the comparisons are put back into a form representation of the total number of entries.

Let us add (not average) the three replications for this case. This reduces our table to a 3^4 layout, or 81 entries. We will assume an SS of 30 here for df $= 16$. The resulting mean square (MS) is 1.87. The table layout is the same as Table 10.15, but with only 1 entry per square (which represents the sum of three).

Our problem is to rank the 4 variables, as well as to determine if there are any interactions between any 2, 3, or 4 of these variables. We need the complete analysis of variance.

We must first look at our 4 main variables (A, B, C, D). Let us summate all the entries of Table 10.15 under A_1; all under A_2; all under A_3. We tabulate — now we are giving representative data — it is not too complicated. See Table 10.16.

Table 10.16   Summation of A Entries

A_1	A_2	A_3
163	−40	39

$n = 81$ items per entry, $N = 243$

3 levels of A give 2df

Total $= 163 - 40 + 39 = 162$

Average $= 54$

$$SS = \frac{1}{81} [163^2 + (-40)^2 + 39^2] - \frac{162^2}{243}$$

$SS = 367 - 108 = 259$

$MS = 130$

Note that in calculating the SS_A we divide by 81 for each level, thus keeping on a total of 243 original data entries.

It looks as though variable A is far from a linear relationship, using summations of levels A_1, A_2, and A_3.

More calculations do the same for variables B, C, and D (not reproduced here).

B entries: 81 items per entry; 2 df
                     $SS = 111$; $MS = 56$

C entries: 81 items per entry; 2 df
                     $SS = 941$; $MS = 471$

D entries: 81 items per entry; 2 df
                     $SS = 1$; $MS = 0.5$

A entries: 81 items per entry; 2 d
                     $SS = 259$; $MS = 130$

Table 10.17   Analysis of Variance (Limited Data)

Effect	*Source*	*SS*	*df*	*MS*
Total, original table	All	1725	242	—
Main factors	A	259	2	130
	B	111	2	56
	C	941	2	471
	D	1	2	0.5
Interaction table after averaging the replication	ABCD	30	16	1.87
Unaccounted for (difference) at this stage of calculation		383	218	1.77

     This tabulation and the SS given make us look for interaction evidence. Certainly factor A is reacting far from a linear relationship with its various levels. Factor D is almost noneffective at its various levels — no change at all to be considered. Factor C goes through the largest range, with its various levels.

     We must start our table of analysis of variance, using only the data we have to this point, as shown in Table 10.17. One statement is necessary here: our "unaccounted for" or "residual" SS value is 383 or greater than the values of SS for three of our main variables. We must account for more of it. Is it interaction?

     Let us see if A and C interact. We build a table (Table 10.18) of A and C, each at three levels, by adding our B_1, B_2, and B_3 values together with our D_1, D_2, and D_3 values also together under their appropriate A, C levels.

     The calculation is corrected for duplication of the individual SS of A and C, using the values $SS_A = 259$, $SS_C = 941$. This calculation also is divided by 27 — the items per entry — to keep an equivalent basis of 243 original data entries. Thus

$$\frac{243 \text{ total entries}}{9 \text{ entries above}} = 27$$

The AC interaction is insignificant with SS = 4.

     Study of the above listing will allow the tabulation of C and D single variable summations. We leave it to you, if interested. Also you can then

Table 10.18 Interaction of A and C, Summating B_1, B_2, B_3, and D_1, D_2, D_3

	A_1			A_2			A_3	
C_1	C_2	C_3	C_1	C_2	C_3	C_1	C_2	C_3
131	29	3	60	-30	-70	82	-8	-35

27 items per entry; df $= 2 \times 2 = 4$

$$SS = \frac{1}{27}[131^2 + 29^2 + \cdots + (-8)^2 + (-35)^2] - \frac{162^2}{243}$$

$$= (1312) - 108 = 1204 \text{ (uncorrected)}$$

But SS for A and SS for C must be subtracted ($SS_A = 259$; $SS_C = 941$)

$$SS = 1204 - 259 - 941 = 4 \text{ corrected.}$$

calculate the SS for these C and D variables separately — as we have already listed. See Table 10.19. The CD interaction is insignificant with SS $= 9$.

The interaction of A and D, shown in Table 10.20, has to wait till all calculations are in before we can determine whether it is significant (see later).

There are six possible two-variable interactions: AB, AC, AD, BC, BD, and CD. Unfortunately, we do not have the B_1, B_2, B_3 data, so we cannot

Table 10.19 Interaction of C and D, Summating A_1, A_2, A_3, and B_1, B_2, B_3

	D_1			D_2			D_3	
C_1	C_2	C_3	C_1	C_2	C_3	C_1	C_2	C_3
96	2	-39	87	-1	-36	90	-10	-27

27 items per entry; df $= 2 \times 2 = 4$

$$SS = \frac{1}{27}[96^2 + 2^2 + \cdots + (-10)^2 + (-27)^2] - \frac{162^2}{243}$$

$$SS = 1059 - 108 = 951 \text{ (uncorrected)}$$

This SS includes SS for $C \times D$, for C and for D.

SS must be corrected for the SS_C, SS_D which are 941 and 1.

Thus SS $= 951 - 941 - 1 = 9$

Table 10.20   Interaction of A and D, Summating B_1, B_2, B_3, and C_1, C_2, C_3

	A_1			A_2			A_3	
D_1	D_2	D_3	D_1	D_2	D_3	D_1	D_2	D_3
64	43	56	-16	-15	-9	11	22	6

27 items per entry; df $= 2 \times 2 = 4$

$$\text{SS} = \frac{1}{27}\,(64^2 + 43^2 + \cdots + 22^2 + 6^2) - \frac{162^2}{243}$$

SS $= 381 - 108 = 274$ uncorrected

But we subtract SS for A and D

SS $= 274 - 259 - 1 = 14$ corrected.

calculate B interaction with A, C, or D. Let us assume they have SS values as follows:

Calculated AC	SS $=$	4	Assumed AB	SS $=$	25
CD	SS $=$	9	BC	SS $=$	16
AD	SS $=$	14	BD	SS $=$	7

Sum of squares are additive. The above six total 75. We have taken by these calculations 75 units out of our "unaccounted-for," or residual, SS.

$$\text{SS} = 383 - 75 = 308$$

We still have a way to go to be happy. So we look for interaction between three variables. There are four possible groupings: ABC, ABD, ACD, and BCD.

Let us try the ADC interaction. Going back on our original data, we get a 27 entry table. It requires some work, but here it is in Table 10.21.

Table 10.21   Interaction ADC, Summating B_1, B_2, B_3

	A_1			A_2			A_3		
	D_1	D_2	D_3	D_1	D_2	D_3	D_1	D_2	D_3
C_1	49	38	44	19	21	20	28	28	26
C_2	15	8	6	-10	-12	-8	-3	3	-8
C_3	0	-3	6	-25	-24	-21	-14	-9	-12
A	64	43	56	-16	-15	-9	11	22	6
D_1	59								
D_2		50							
D_3			53						

27 entries; 9 items per entry: df $= 2 \times 2 \times 2 = 8$

$$SS = \frac{1}{9} [49^2 + 15^2 + \cdots + (-8)^2 + (-12)^2] - \frac{162^2}{243}$$

SS $= 1339 - 108 = 1231$ uncorrected.

The correction here is as follows:

$SS_A = 259$	$SS_{AC} = 4$
$SS_C = 941$	$SS_{AD} = 14$
$SS_D = 1$	$SS_{CD} = 9$

Total SS corrected $= 1228$

SS $= 1231 - 1228$ corrected

SS $= 3$ corrected

This SS tabulation includes all of these corrections plus the *ADC* interaction we are wanting.

Note: If you summate the C_1, C_2, C_3 here, you get our earlier table on *AD* interactions. You now see how the data lock together.

We are not getting anywhere fast.

We cannot calculate any three-variable interactions involving *B* for which the data have not been presented. We will assume the SS for these to be

ABC	SS $= 24$
ABD	SS $= 8$
BCD	SS $= 11$
plus *ACD*	SS $= 3$
	total $= 46$

Now we set up our final analysis of variance, Table 10.22.

ANALYSIS OF THE INTERACTION VARIANCE. In Table 10.22 we add all of the sum of squares of differences and subtract from our total individual SS. This gives us an "original residual" of 262 SS. We have accounted for 85% of our spread of values as follows:

Main variables	1312	SS values
Two-variable interactions	75	SS values
Three-variable interactions	46	SS values
Four-variable interactions	30	SS values
	1463	
Total calculated	1725	
Then residual is	262	

Table 10.22 Analysis of Variance for Study of Interaction; * indicates significant while # means not significant.

Source of Variation	Factors	SS	df	MS	F Ratio
Total individual	All	1725	242	—	—
Main factors	A	259	2	130	80.0*
	B	111	2	56	34.5*
	C	941	2	471	312.0*
	D	1	2	0.5	— #
Interaction — two factors	AB	25	4	6.25	3.67*
(two-variable interactions)	AC	4	4	1.0	— #
	AD	14	4	3.50	2.16#
	BC	16	4	4.0	2.37?
	BD	7	4	1.75	— #
	CD	9	4	2.25	— #
Interaction — three factors	ABC	24	8	3.0	1.92#
(three-variable interactions)	ABD	8	8	1.0	— #
	ACD	3	8	0.38	— #
	BCD	11	8	1.38	— #
Interaction — four factors (four-variable interactions)	ABCD	30	16	1.87	— #
Original residual		262	162	1.62	—
Revised residual including all 3 and 4 variable interactions		338	210	1.61	—
Total		1725	242	—	—

Putting these on the same basis for the degrees of freedom gives the mean square (MS) in the next to last column of Table 10.22. Now we divide these values from MS by the MS of the residual (1.62) and get the F ratio. If these values are in the realm of slightly above 1 or lower than 1, we know they are not significant. So we can write n.s. (not significant) opposite many of our entries. If they are going to be very high, we know they are significant — but we calculate it. (These general statements come from a view of Table 10.22.)

We start from the bottom of the table and work up.

$$\text{For } ABCD, F = \frac{1.87}{1.62} = 1.15 \qquad \begin{array}{l} \text{df numerator} = 16 \\ \text{df denominator} = 162 \end{array}$$

Read Table 10.3 variance ratios for say 15 over ∞. The 1.15 does not even exceed the value for 0.90 confidence. Thus, *no significance*.

$$\text{For } ABC, F = \frac{3.00}{1.62} = 1.92 \qquad \frac{\text{df numerator} = 8}{\text{df denominator} = 162}$$

Table 10.3 = F ratio for 8 over ∞

$$0.90 \; = 1.67 \qquad 0.95 \; = 1.94 \qquad 0.975 = 2.19$$

The 1.92 value approaches the 0.95 table value. Significance approaches 0.95 but this value is the lowest we generally accept as significant when using the F test. Thus there is questionable significance.

OTHER 3-VARIABLES. All are much below the ABC above discussed, so *no significance*.

AN INTERIM LOOK. Before studying the two-variable interactions, we look at our 3-variable and 4-variable data. We see that it should, by common sense and by no indication of significance, be lumped with our overall experimental error — that is, into the residual. This gives us a new residual *and a new divisor* for use in the two-variable and single evaluations. We do this as follows:

Original	SS = 262	df = 162
ABCD	= 30	= 16
ABC	= 24	= 8
ABD	= 8	= 8
ACD	= 3	= 8
BCD	= 11	= 8
Totals	338	210

This gives the revised residual. Our new divisor is $338/210 = 1.61$, which compares to the original divisor of 1.62. This proximity of value is further proof that the 3-variable and 4-variable interactions were nothing more than experimental error (that is as a group).

TWO-VARIABLE INTERACTIONS. Using our new divisor, we calculate F ratios and compare to table values, thus forming Table 10.23, where all

$$\frac{\text{df} = 4}{\text{df} = 210}$$

We can say that interaction of AB is very significant; interaction of BC is borderline but acceptable. Others show no interaction.

Table 10.23   Two-Variable Interactions

	F		from Table 10.3
AB	3.67	←	0.90 = 1.94
AC	—	Do these	0.95 = 2.37
AD	2.16		0.975 = 2.79
BC	2.37	exceed these?	0.99 = 3.32
BD	—	→	0.995 = 3.72
CD	—		

For our main variables, we calculate *F* as

80.0 for *A*   34.5 for *B*   312.0 for *C*   n.s. for *D*

$$\text{All } \frac{df = 2}{df = 210}$$

Highest table value is 5.30 for 0.995 confidence. Variables *A*, *B*, and *C* are highly significant. Variable *D* is of no importance.

Experimental error is given by the SS of the revised residual 338 divided by the df of 210. This MS value of 1.61 is a measure of the error related to the MS of the various other facets of the table.

SUMMARY OF STEPS FOR ANALYSIS OF VARIANCE.   Looking back now, you can see the steps to take in this work.

Step 1. By coding, attempt to get all original data on as simple a numerical basis as possible by (1) addition or subtraction of a constant, or (2) multiplication or division of a constant. Whichever you do, the *F* ratios will not be affected. Getting into some negative and some positive values does not matter (they square up to positive values).

Step 2. Calculate the total sum of squares for the individual data. The df here is $n - 1$ for *n* data entries.

Step 3. Calculate the SS for all levels of each variable. These will be all individually grouped columns, all individually grouped lines, *and* other identified grouped variables in the tabulation. These SS give the basis for the effects of the main factors, or the main variables. These are known as *between-level* variations. Determine the df for these cases. Three levels of a variable give $(3 - 1)$ df.

Step 4. Calculate the SS for interaction between variables. Here we use the summation of observations for each of the combinations of levels for the variables. But there is a correction to be applied since the calculation also includes the SS for each of the included variables — the SS which has already been calculated in Step 3 above. So we subtract these to get the interaction SS. For instance, calculation of SS for variable AB includes the SS for A and the SS for B. So

$$SS_{\text{interaction } AB} = SS_{\text{calculated}\atop\text{as above}} - (SS_A + SS_B)_{\text{calculated}\atop\text{as in Step 3}}$$

For instance, calculation of SS for interaction ABC includes the SS for AB, BC, AC, A, B, and C. All must be determined separately and subtracted. So

$$SS_{\text{interaction } ABC} = SS_{\text{calculated}\atop\text{as above}}$$

$$- (SS_{AB} + SS_{AC} + SS_{BC})_{\text{calculated as}\atop\text{in earlier part}\atop\text{of Step 4}}$$

$$- (SS_A + SS_B + SS_C)_{\text{calculated}\atop\text{as in Step 3}}$$

The df for interaction is the product of the individual df values making up that interaction.

Step 5. Total the SS from Steps 3 and 4, then subtract from the SS from Step 2, and get the SS for error, or "residual" or "inherent" variation, or "within group" variation.

Step 6. Tabulate an analysis of variance table. Calculate the SS/df for each condition and arrive at the MS (mean squares) for all conditions except the total.

Step 7. Calculate the F ratios, which are

$$\frac{MS}{MS_{\text{residual}}}$$

for each interaction of 3 or more variables.

Step 8. Compare these F ratios to the table F ratios determined for the proper

$$\frac{\text{df numerator}}{\text{df denominator}}$$

basis for the various confidences.

Step 9. A confidence of

0.9	is not considered important — there is some question
0.95	is acceptable — but the lowest we generally accept
>0.95	is considered positive evidence
0.995	is very positive evidence

Step 10. If the higher order interactions are not significant, total their SS and df values and add to the "residual" values. Use this new residual

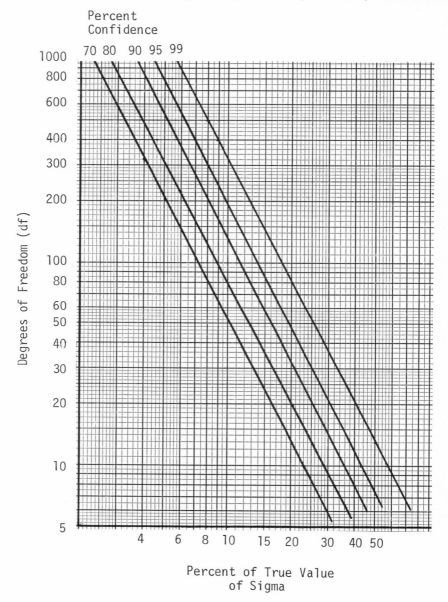

Figure 10.4 Sample size (df + 1) for estimate of standard deviation to within stated percentage of true value, when sample size is n, $(n - 1)$ = df. Adapted from Reference 26, M. G. Natrella, Experimental Statistics, *Natl. Bur. of Std. Handbook*, 91, Aug 1, 1963.

value for evaluation of two-variable and one-variable F ratios and re-evaluate, etc (Steps 7, 8, 9).

Looking back: The dividing up of the total sum of squares (SS) for a group of data is an attempt to allocate the whole variable of that data into segments — effects of main variables, effects of interaction, the residual or perhaps experimental error, etc. But do not forget that if interaction is not checked for, it falls into this residual grouping. The concept of getting the MS for the residual down to a level of complete insignificance seems one to be on the watch for.

10.6 SAMPLE SIZE FOR ESTIMATION OF VARIANCE AND SIGMA

From the examples we have used, we soon realize that estimations of σ from a limited sample size are not very precise. While averages from a rather limited set of data are readily useful, we require a much larger sample size to get an equal estimate of σ. In the first case, we are estimating only a 50% point on the cumulative probability paper plot. In the latter case we are trying to plot the whole curve. The conclusion is therefore expected.

Sometimes it is desirable to know the sample size required to estimate σ to a desired confidence. Figure 10.4 gives the degrees of freedom —thus the sample size — for 70% to 99% confidence and for a range from 5% to 50% of the true value of σ.

EXAMPLE E10-8.   If we want to be within 10% of true value of σ, with 90% confidence, we must have a sample size of approximately 130 to 140 units.

Truly, confidence in σ comes very expensive.

11

COMPARISON OF TWO SETS WITH DEFECTIVES FOR SIGNIFICANT DIFFERENCES
Comparison with Attributes

The only significant difference
between a beautiful woman and an
ugly woman is the way they look.

So far, we have been concerned with data which were expressed in measurable fashion and were defined in Section 1.2 as variables. When we get into a class of defectives which are limited to go-no-go classifications, we no longer have a mass of data to estimate our average and sigma. That is, a group of 50 bottles would give us 50 measurements on out-of-round of the finish. These could be averaged and a standard deviation could be calculated. These cases we have covered.

When, however, we get over into attributes instead of variables, we end up perhaps with only two figures — those that "go" and those that "do not go" or those that are acceptable in quality and those that are not. We may have 5 defectives in a 50 sample. How do we use this in a statistical fashion? For instance, suppose we have two sets as in Table 11.1. Is there a significant difference in these two sets, or is it a chance sampling variation?

Table 11.1   Two Sets of Attributes

	Number Defective	Number Good	Total Number Inspected
Set 1	5	120	125
Set 2	7	43	50
Total	12	163	175

171

We will now discuss this type of comparison, using these formulas for attributes:

$$\sigma_{p(1)} = \left[\frac{(p_1)(1 - p_1)}{n}\right]^{\frac{1}{2}}$$

where p_1 = fraction defective of Set 1 and n = number of samples in p_1 group

$$\sigma_{p(2)} = \left[\frac{(p_2)(1 - p_2)}{m}\right]^{\frac{1}{2}}$$

where m = number of samples in p_2 group

$$p' = \frac{\sum (\text{defectives in two groups})}{\sum (\text{samples in two groups})}$$

$$p' = \frac{p_1 n + p_2 m}{n + m}$$

$$\sigma_{p(1)-p(2)} = \left[(p')(1 - p')\left(\frac{1}{n} + \frac{1}{m}\right)\right]^{\frac{1}{2}}$$

For comparison:

$$Z = \frac{(p_2 - p_1)}{\sigma_{p(2)-p(1)}}$$

where Z = standard normal variable. (See formulas with Table 11.4).

11.1   BY GRAPHICAL PROCEDURES

A very useful chart for significance when we are using attributes is given by Li (Ref. 23). Figure 11.1 represents the average case *when the two samples to be compared are of equal or very nearly equal size (within 3% or so).*

In such a case, a straight line is drawn on the chart connecting the sample size ($n_1 = n_2$) on the n-scale to the difference between the number of defectives ($D = d_2 - d_1$) on the D-scale. The two S scales are then found so that the sum of the number of defectives ($S = d_2 + d_1$) appears just above the straight line on the upper S-scale but just below it on the lower. These two S-scales identify the region of significance. If the $D - n$ line does not cut all the S-scales, the least statement of significance that can be made is given by the first significance region (from the bottom of the chart) through which the line passes.

EXAMPLE E11-1.   With 4 defectives in 20, 1 defective in 20: difference not significant at any level.

EXAMPLE E11-2.   With 6 defectives in 30, 1 defective in 30: doubtful significance between 0.05 and 0.10 probability.

EXAMPLE E11-3.   With 9 defectives in 40, 2 defectives in 40: definitely significant, between 0.01 and 0.05 probability.

EXAMPLE E11-4.   With 11 defectives in 50, 2 defectives in 50: very significant, between 0.001 and 0.01 probability.
Note the effect of sample size.

11.2 SAMPLE SIZE

Figure 11.1 can be used in a reverse fashion to determine if the sample size is large enough. The example facing Figure 11.1 is typical for this usage.

11.3 AVERAGE, SIGMA, AND RANGE FOR AVERAGE FOR ATTRIBUTES

We take a large random sample (n) from the population, and count (c) the defective ones. These counts must not necessarily be defective. They may represent any characteristic of interest to us, good or bad. This count is the proportion, *expressed as a fraction* (0 to 1),

$$p = \frac{c}{n}$$

where c = number with the characteristic of interest and n = total lot sample.

EXAMPLE E11-5.   9 samples are defective from a group lot of 50:

$$p_1 = \frac{9}{50} = 0.18$$

$$\sigma_{p(1)} = \left[\frac{(p_1)(1 - p_1)}{n}\right]^{\frac{1}{2}}$$

$$= \left[\frac{(0.18)(1 - 0.18)}{50}\right]^{\frac{1}{2}} = \left[\frac{(0.18)(0.82)}{50}\right]^{\frac{1}{2}} = (0.00295)^{\frac{1}{2}} = 0.054$$

For 5 counts in a lot of 50, we have

$$p_2 = \frac{5}{50} = 0.10 \text{ and } \sigma_{p(2)} = 0.042$$

What are the ranges for p_1 and p_2 to be expected? We can read the

How to Use Figure 11.1 for Significant Difference and Sample Size for Attributes

CAUTION.   Do not use this chart (Figure 11.1) with a sample size of eight or less.

Do not use this chart (Figure 11.1) where the percentage of defectives is below 1% of the total population.

SIGNIFICANT DIFFERENCE.   Draw a straight line connecting sample size (for each group) from the n-scale to the difference between the number of defectives on the D-scale. *The sample sizes must be the same or within 3% of each other.* The two S-scales are then found so that the sum of the number of defectives appears *just above* the straight line on the upper S-scale and *just below* it on the lower S-scale. These two S-scales define the region of significance. If the D-to-n line does not cut all the S-scales, the least statement of significance that can be made is given by the first significance region (from the bottom of the chart) through which the line passes.

Take the case   4 defectives in 20
                   1 defective in 20   difference not significant at any level
Take the case   11 defectives in 50
                   2 defectives in 50   difference very significant (between P 0.001 and 0.01).

SAMPLE SIZE.   This determination is reverse of above. Take the case of difference in number of defectives as 7, and sum of number of defectives as 9. Draw line from 7 on the D-scale through slightly below 9 on the P 0.01 scale and read off sample number of 18 required on n-scale.

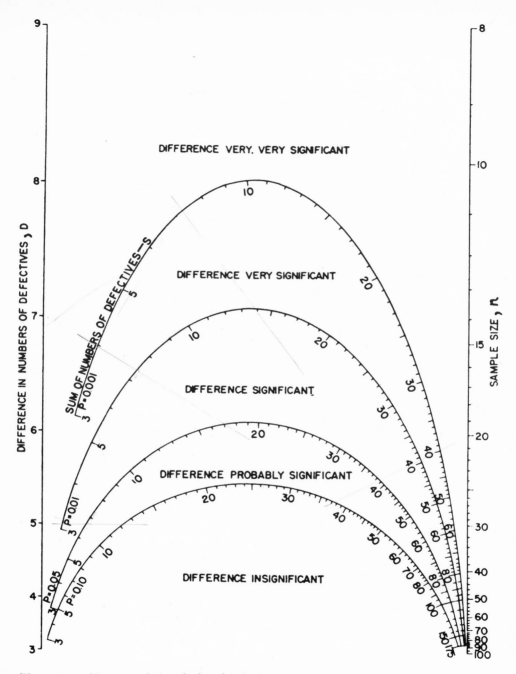

Figure 11.1 Nomograph to calculate (a) significance for attributes and (b) sample size required for definite significance. From Reference 23, C. H. Li, "A Nomograph for Evaluating the Significance of Some Test Results," *A.S.T.M. Bull.* Dec. 1953 (Issue 194) T.P. 218, by permission of American Society for Testing and Materials. Copyright 1953 by American Society for Testing and Materials.

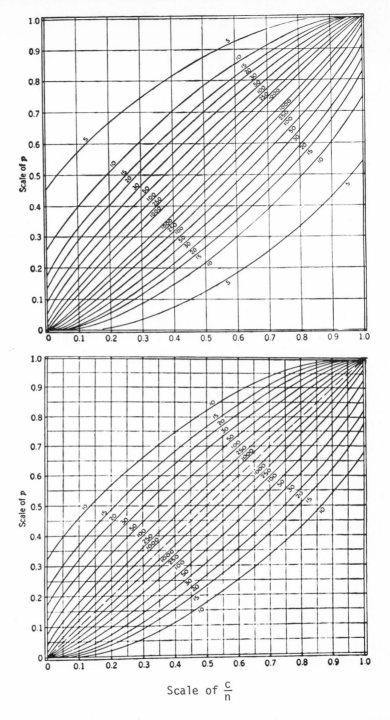

Figure 11.2 Charts for confidence limits for averages for limits; (a) confidence coefficient = 0.90, (b) confidence coefficient = 0.95; numbers along line indicate sample size, *n*. Based on Binomial from Reference 13, W. J. Dixon and F. J. Massey, Jr., *Introduction to Statistical Analysis*, 2d ed., McGraw-Hill, 1957, chart (a) used with permission of McGraw-Hill Book Company. Copyright 1957 by McGraw-Hill Book Company. Chart (b) used with permission of E. S. Pearson on behalf of Biometrika Trust. Copyright 1934 by Biometrika Trust.

ranges from charts given in Figure 11.2. These are based on the binomial (see Section 12).

For 90% confidence (Fig. 11.2a) the ranges are as follows for Example E11-5:

$$p_1 = 0.18 \qquad \frac{c}{n} = 0.18 \qquad p_2 = 0.10; \frac{c}{n} = 0.10$$

$$\text{Range } p_1 \text{ 0.09 to 0.30} \qquad \text{Range } p_2 \text{ 0.04 to 0.20}$$

These two p values overlap greatly.

We can also read off the number of samples required for various narrower ranges of p_1 and p_2 from Figure 11.2.

Table 11.2   Ranges for Attributes from Figure 11.2

n	$p_1 = \left(\dfrac{c}{n} = 0.18\right)$	$p_2 = \left(\dfrac{c}{n} = 0.10\right)$
50	0.09 — 0.30	0.04 — 0.20
100	0.13 — 0.26	0.06 — 0.17
250	0.15 — 0.24	0.05 — 0.14

Note: It takes a sample size of 250 to give a definite nonoverlap of the p_1 and p_2 values.

11.4   TESTS FOR SIGNIFICANT DIFFERENCES FOR ATTRIBUTES USING LARGE SAMPLE WHERE THE PRODUCT OF SAMPLE SIZE (n) AND PERCENT DEFECTIVE (p) IS AT LEAST 5 OR MORE

EXAMPLE E11-6.   Take the comparison shown in Table 11.3, and use the formula

$$\sigma_{p(1)-p(2)} = \left[(p')(1 - p')\left(\frac{1}{n_1} + \frac{1}{n_2}\right)\right]^{\frac{1}{2}}$$

Table 11.3   Comparison for Attributes — Example E11-6

	Number Defective (np)	Number Good ($n - np$)	Total Number Inspected (n)
Set 1	5	120	125
Set 2	7	43	50
Total	12	163	175

where

p_1 = fraction defective of set 1 = 5/125

p_2 = fraction defective of set 2 = 7/50

    Note: for *both* cases:

$$np = \frac{(125)\ 5}{(125)} = 5 \quad \text{and} \quad np = \frac{(50)\ 7}{(50)} = 7$$

p' = fraction defective of two sets combined

$$p' = \frac{5 + 7}{125 + 50} = \frac{12}{175} = 0.07$$

n_1 = number of samples in set 1

n_2 = number of samples in set 2

Calculating, we get

$$\sigma_{p(2)-p(2)} = \left[(0.07)(0.93)\left(\frac{1}{125} + \frac{1}{50} \right) \right]^{\frac{1}{2}}$$

$$= [(0.0651)(0.008 + 0.02)]^{\frac{1}{2}} = (0.00182)^{\frac{1}{2}} = 0.0427$$

Take

$$Z = \frac{p_2 - p_1}{\sigma_{p(2)-p(1)}} = \frac{0.14 - 0.04}{0.0427} = 2.34$$

The quantity Z is approximately normally distributed for large samples — where (np) is equal to or over 5. The critical values for Z for various levels of confidences are found in Table 11.4.

Our Z value is 2.34, or there is acceptable significant difference between sets 1 and 2.

How to Use Table 11.4

Use the formula

$$\sigma_{p(2)-p(1)} = \left[(p')(1 - p')\left(\frac{1}{n_1} + \frac{1}{n_2} \right) \right]^{\frac{1}{2}}$$

where

p_1 = fraction defective of set 1
p_2 = fraction defective of set 2
p' = fraction defective of two sets combined
n_1 = number of samples in set 1
n_2 = number of samples in set 2

Then calculate Z (the standard normal variable)

$$Z = \frac{(p_2 - p_1)}{\sigma_{p(2)-p(1)}}$$

Compare Z with table value to get confidence value, or range. The confidence is based on the following:

very positive confidence 0.99 value exceeded
acceptable confidence 0.95 value exceeded
confidence has some question 0.90 value exceeded

Table 11.4 Confidence Level Relation to Critical Values Useful for Comparing Significant Differences of Attributes, Based on normal distribution and requiring sample size (n) times fraction defective (p) equal to or greater than 5. *Note:* If the calculated value Z from the data is greater than the Table Z value, then the two proportions are significantly different from the table confidence level.

Two-Tailed Confidence Level	Critical Value (Z)
70%	1.036
80	1.282
90	1.645
95	1.960
97.5	2.241
98	2.326
99	2.576
99.5	2.807
99.8	3.090
99.9	3.291

EXAMPLE E11-7. Suppose the data are as given in Table 11.5.

Table 11.5 Comparison for Attributes — Example E11-7

	Number Defective (np)	Number Good ($n - np$)	Total Number Inspected (n)
Set 3	5	45	50
Set 4	9	41	50
Total	14	86	100

Calculate:

$$\sigma_{p(4)-p(3)} = \left[(0.14)(0.86)\left(\frac{1}{50} + \frac{1}{50}\right)\right]^{\frac{1}{2}}$$

$$= [(0.12)(0.04)]^{\frac{1}{2}}$$

$$= (0.0048)^{\frac{1}{2}} = 0.069$$

$$Z = \frac{(0.18 - 0.10)}{0.069} = \frac{0.08}{0.069} = 1.16$$

This is not a significant difference (compare Table 11.4).

EXAMPLE E11-8. Take the comparison shown in Table 11.6.

Table 11.6   Comparison for Attributes — Example E11-8

	Number Defective	Number Good	Total Number Inspected
Set 1	11	39	50
Set 2	5	45	50
Total	16	84	100

Calculate:

$$p_1 = 11/50 = 0.22 \qquad p_2 = 5/50 = 0.10 \qquad p' = 16/100 = 0.16$$

$$\sigma_{p(1)-p(2)} = \left[(0.16)(0.84)\left(\frac{1}{50} + \frac{1}{50}\right)\right]^{\frac{1}{2}}$$

$$= [(0.134)(0.04)]^{\frac{1}{2}} = (0.00536)^{\frac{1}{2}} = 0.073$$

$$Z = \frac{0.22 - 0.10}{0.073} = \frac{0.12}{0.073} = 1.64$$

$$= 1.64 \text{ at } 90\% \text{ confidence level}$$

Therefore difference has some question of significance.

EXAMPLE E11-9. Take the comparison of data in Table 11.7.

Table 11.7   Comparison for Attributes — Example E11-9

	Number Defective	Number Good	Total Number Inspected
Set 1	9	31	40
Set 2	5	95	100
Total	14	126	140

Calculate:

$$p_1 = 9/40 = 0.22 \qquad p_2 = 5/100 = 0.05 \qquad p' = 14/140 = 0.10$$

$$\sigma_{p(2)-p(1)} = \left[(0.10)(0.90)\left(\frac{1}{40} + \frac{1}{100}\right) \right]^{\frac{1}{2}}$$

$$= [(0.09)(0.025 + 0.010)]^{\frac{1}{2}} = (0.003)^{\frac{1}{2}} = 0.055$$

$$Z = \frac{(0.22 - 0.05)}{(0.055)} = \frac{(0.17)}{(0.055)} = 3.10$$

Therefore, difference is significant to over 0.99 confidence.

11.5   THE CHI-SQUARE (χ^2)

Chi-square is a pile-up or summation of the data in relation to what the expected (theoretical) data would be. The comparison of the expected or theoretical (E) frequency for a given lot is made to the observed (O) frequency.

The chi-square test can be used beyond a two-category classification (like go-no-go) for comparison of several classifications. We might have three classifications, such as acceptable, reworkable, or unusable; or perhaps, size, color, and defective. We have four problems in the usage of chi-square.

— A decision on the (E) expected value. It can be the theoretical, the expected, the desired, the average of the tests made, or a commonsense value based on an analysis of the observed data, etc. References 13 and 26 give some examples.

— The values in each cell should be at least 5 from a rather large distribution. To attain this value, we may combine several cells.

— A factor called the "correction for continuity" is generally recommended to be used where the df is one — that is, in a 2 × 2 table. For comparison of two sets of defectives, this df applies. It may not be at all important if the sample size is large. But it is better to be safe and use it for all cases of df = 1. It *does not apply* for cases where df is 2 or more.

— The determination of the degrees of freedom may seem cumbersome. Here it has nothing to do with sample size but rather covers the tabular layout required for the summarized data. Its best definition and determination are in Section 9.5. Remember, for comparing two sets of defectives, the df is always 1.

EXAMPLE E11-10, *with Two Sets of Defectives.*   Take the case of two equal lots, which is the simplest to illustrate, and refer to Table 11.8 in

Table 11.8 Data for Example E11-10

	Number Defective	Number Good
Set 1	$5(-O_1)$	$45(-O_2)$
Set 2	$10(-O_3)$	$40(-O_4)$
Total	15	85

which columns 2 and 3 give four observed O values. Assume both lots are equal in quality — this is the expectation! — or $(5 + 10)/2$

Set 1	$7.5(-E_1)$	$42.5(-E_2)$
Set 2	$7.5(-E_3)$	$42.5(-E_4)$

Columns 2 and 3 give four expected E values. Example E11-16 later herein gives more details on determining expected value (E). Use the formula:

$$\chi^2 = \frac{(O_1 - E_1)^2}{E_1} + \frac{(O_3 - E_3)^2}{E_3} + \frac{(O_2 - E_2)^2}{E_2} + \frac{(O_4 - E_4)^2}{E_4}$$

$$= \frac{(5 - 7.5)^2}{7.5} + \frac{(10 - 7.5)^2}{7.5} + \frac{(45 - 42.5)^2}{42.5} + \frac{(40 - 42.5)^2}{42.5}$$

$$= \frac{6.25}{7.5} + \frac{6.25}{7.5} + \frac{6.25}{42.5} + \frac{6.25}{42.5} = 1.86$$

SUMMARY.   Briefly, since the first figure in the numerator is the actual and the second is the expected, we get as a general formula:

$$\chi^2 = \text{summation of} \left[\frac{(\text{observed} - \text{expected})^2}{\text{expected}} \right]$$

or $$\chi^2 = \Sigma \left[\frac{(O - E)^2}{E} \right]$$

where O = observed frequency; E = expected frequency; and Σ = summation over all the separate classes (but not the average column at the end of the tabulation).

INTERPRETATION BY MEANS OF TABLE 11.9.   Values of χ^2 for 1 df are:

$P = 0.50$	$\chi^2 = 0.455$	$P = 0.10$	$\chi^2 = 2.71$
$= 0.25$	$= 1.32$	$= 0.05$	3.84 our value $= 1.86$

Thus, there are between 10 and 25 chances in 100 that the differences observed could have arisen solely due to chance. Using a 10% (0.10) level of $\chi^2 = 2.71$, we could not establish a difference from these two sets of data. The value 2.71 must be exceeded to establish significant difference at the 10% risk level.

Table 11.9 Values of Chi-Square (χ^2). F is Probability that Value in Table will be Exceeded, for Given Number of Degrees of Freedom n (from Ref. 16)*

n	Values of F												
	0.995	0.990	0.975	0.950	0.900	0.750	0.500	0.250	0.100	0.050	0.025	0.010	0.005
1	0.0^4393	0.0^3157	0.0^3982	0.00393	0.0158	0.102	0.455	1.32	2.71	3.84	5.02	6.63	7.88
2	0.0100	0.0201	0.0506	0.103	0.211	0.575	1.39	2.77	4.61	5.99	7.38	9.21	10.6
3	0.0717	0.115	0.216	0.352	0.584	1.21	2.37	4.11	6.25	7.81	9.35	11.3	12.8
4	0.207	0.297	0.484	0.711	1.06	1.92	3.36	5.39	7.78	9.49	11.1	13.3	14.9
5	0.412	0.554	0.831	1.15	1.61	2.67	4.35	6.63	9.24	11.1	12.8	15.1	16.7
6	0.676	0.872	1.24	1.64	2.20	3.45	5.35	7.84	10.6	12.6	14.4	16.8	18.5
7	0.989	1.24	1.69	2.17	2.83	4.25	6.35	9.04	12.0	14.1	16.0	18.5	20.3
8	1.34	1.65	2.18	2.73	3.49	5.07	7.34	10.2	13.4	15.5	17.5	20.1	22.0
9	1.73	2.09	2.70	3.33	4.17	5.90	8.34	11.4	14.7	16.9	19.0	21.7	23.6
10	2.16	2.56	3.25	3.94	4.87	6.74	9.34	12.5	16.0	18.3	20.5	23.2	25.2
11	2.60	3.05	3.82	4.57	5.58	7.58	10.3	13.7	17.3	19.7	21.9	24.7	26.8
12	3.07	3.57	4.40	5.23	6.30	8.44	11.3	14.8	18.5	21.0	23.3	26.2	28.3
13	3.57	4.11	5.01	5.89	7.04	9.30	12.3	16.0	19.8	22.4	24.7	27.7	29.8
14	4.07	4.66	5.63	6.57	7.79	10.2	13.3	17.1	21.1	23.7	26.1	29.1	31.3
15	4.60	5.23	6.26	7.26	8.55	11.0	14.3	18.2	22.3	25.0	27.5	30.6	32.8
16	5.14	5.81	6.91	7.96	9.31	11.9	15.3	19.4	23.5	26.3	28.8	32.0	34.3
17	5.70	6.41	7.56	8.67	10.1	12.8	16.3	20.5	24.8	27.6	30.2	33.4	35.7
18	6.26	7.01	8.23	9.39	10.9	13.7	17.3	21.6	26.0	28.9	31.5	34.8	37.2
19	6.84	7.63	8.91	10.1	11.7	14.6	18.3	22.7	27.2	30.1	32.9	36.2	38.6
20	7.43	8.26	9.59	10.9	12.4	15.5	19.3	23.8	28.4	31.4	34.2	37.6	40.0
21	8.03	8.90	10.3	11.6	13.2	16.3	20.3	24.9	29.6	32.7	35.5	38.9	41.4
22	8.64	9.54	11.0	12.3	14.0	17.2	21.3	26.0	30.8	33.9	36.8	40.3	42.8
23	9.26	10.2	11.7	13.1	14.8	18.1	22.3	27.1	32.0	35.2	38.1	41.6	44.2
24	9.89	10.9	12.4	13.8	15.7	19.0	23.3	28.2	33.2	36.4	39.4	43.0	45.6
25	10.5	11.5	13.1	14.6	16.5	19.9	24.3	29.3	34.4	37.7	40.6	44.3	46.9
26	11.2	12.2	13.8	15.4	17.3	20.8	25.3	30.4	35.6	38.9	41.9	45.6	48.3
27	11.8	12.9	14.6	16.2	18.1	21.7	26.3	31.5	36.7	40.1	43.2	47.0	49.6
28	12.5	13.6	15.3	16.9	18.9	22.7	27.3	32.6	37.9	41.3	44.5	48.3	51.0
29	13.1	14.3	16.0	17.7	19.8	23.6	28.3	33.7	39.1	42.6	45.7	49.6	52.3
30	13.8	15.0	16.8	18.5	20.6	24.5	29.3	34.8	40.3	43.8	47.0	50.9	53.7

*Columns 0.995, 0.975, 0.025, and 0.005 are abridged with permission from Catherine M. Thompson, "Table of Percentage Points of the χ^2 Distribution," *Biometrika*, Vol. XXXII, Part II (1941), pp. 188–89. The remainder of the table is abridged from Table IV of Fisher & Yates: *Statistical Tables for Biological, Agricultural and Medical Research*, published by Oliver & Boyd, Edinburgh, and by permission of the authors and publishers.

Formulas for Usage of Chi-square (χ^2), Table 11.9

The chi-square test is applied to cases of *attributes* as observed (O) versus attributes expected (E). Chi-square $= \chi^2$

$$\chi^2 = \sum \left[\frac{(O - E)^2}{E} \right]$$

where O is observed frequency and E is expected frequency. Chi-square is the summation, therefore

$$\chi^2 = \frac{(O_1 - E_1)^2}{E_1} + \frac{(O_3 - E_3)^2}{E_3} + \frac{(O_2 - E_2)^2}{E_2} + \frac{(O_4 - E_4)^2}{E_4}$$

The *observed* block of data is given in Table 11.10. It is compared with the *expected* block of data which is shown in Table 11.11. Let $a = O_1$, $b = O_2$, $c = O_3$, $d = O_4$, and the block is as shown in Table 11.12. Then we get

$$\chi^2 = \frac{(ad - bc)^2 \times N}{(a + b)(b + d)(c + d)(a + c)}$$

without the continuity correction, where N is the total number of samples. *Correcting for continuity*, we get

$$\chi^2 = \frac{[ad - bc - (N/2)]^2 \times N}{(a + b)(b + d)(c + d)(a + c)}$$

In using Table 11.9 it is necessary to remember that, if the calculated chi-square value is greater than the table value, then there is evidence of a difference between the two sets of data. The confidence of the decision is based on the chi-square table, as follows:

0.01 probability exceeded, very positive confidence
0.05 probability exceeded, confidence is acceptable
0.10 probability exceeded, confidence has some question

See text for examples for corrections applied, and method of using formulas.
Note: The degrees of freedom for *comparing two lots* of fraction defective is only 1. The first line of the table is therefore our main interest. Later in this section there is further discussion on the calculation of the degrees of freedom.

Table 11.10 Observed Block of Data

	Number Defective	Number Good
Set 1	O_1	O_2
Set 2	O_3	O_4

Table 11.11 Expected Block of Data

	Number Defective	Number Good
Set 1	E_1	E_2
Set 2	E_3	E_4

Table 11.12 Block for χ^2 Formula

	Number Defective	Number Good
Set 1	a	b
Set 2	c	d

CORRECTION FOR CONTINUITY.   This is an arbitrary correction based on the fact that our actual distribution is discrete, while the function χ^2 is a continuous one. We use the formula

$$\chi^2 = \frac{[ad - bc - (N/2)]^2 \times N}{(a+b)(b+d)(c+d)(a+c)}$$

where a, b, c, d, are indicated in tabulation (Table 11.13). Corrected for continuity we get

$$\chi^2 = \frac{[(5 \times 43) - (120)(7) - (175/2)]^2 \times 175}{(5+120)(120+43)(7+43)(5+7)}$$

$$= \frac{(215 - 840 - 87)^2 \times 175}{(125)(163)(50)(12)} = \frac{(-712)^2 \times 175}{(125)(163)(600)} = 7.2$$

Our value of 7.2 falls better than 0.01 probability. Thus there is a very positive significant difference between the two sets of data.

Table 11.13   Data for Correction for Continuity

	Number Defective	Number Good	Total Number Inspected
Set 1	5 (a)	120 (b)	125
Set 2	7 (c)	43 (d)	50
Total	12	163	175

EXAMPLE E11-11, *where Only Two Possibilities Exist such as Tossing a Coin.*   The 1×2 case (the df for χ^2 is 1) is illustrated in tossing a coin; only heads or tails can result. Suppose we have 60 heads and 40 tails. The expected is 50 heads and 50 tails. Therefore

$$\chi^2 = \frac{(60 - 50)^2}{50} + \frac{(40 - 50)^2}{50} = 4.0$$

For df 1, and 0.01 probability, we get a table value of $\chi^2 = 6.63$. Conclusion: We cannot say the coin is biased.

EXAMPLE E11-12, *More samples on coin tossing.*   We toss a coin and get a cumulative total of 60 tails and 80 heads, in 140 tosses.

$$\chi^2 = \frac{(60 - 70)^2}{70} + \frac{(80 - 70)^2}{70} = \frac{100}{70} + \frac{100}{70} = 2.857$$

Under Table 11.9 we see that this value falls near the 10% probability. We begin to think the coin is biased, but full proof awaits still more tosses of it.

EXAMPLE E11-13, *Do the Coins Differ?* Now let us compare the two cases of Examples E11-11 and E11-12. Are they suspected to be different coins? The table is

Coin (1)?   60(a) heads +   40(b) tails = 100
Coin (2)?   60(c) heads +   80(d) tails = 140
Total   120   + 120   = 240

$$\chi^2 = \frac{[(60 \times 80) - (60 \times 40) - (240/2)]^2 240}{(60 + 40)(40 + 80)(60 + 80)(60 + 60)}$$

$$= \frac{(2280)^2 240}{(100)(120)(140)(120)} = 6.2$$

This is between 0.98 and 0.99 probability of showing a difference between the two coins.

EXAMPLE E11-14, *Frequency of Accidents, the $1 \times n$ case, the df for χ^2 being $(n - 1)$.* Take accidents in a particular period related to month as follows:

December, 13 accidents; May, 9 accidents; September, 5 accidents

Expected (E) is

$$\frac{5 + 9 + 13}{3} = 9$$

$$\chi^2 = \frac{(5 - 9)^2}{9} + \frac{(9 - 9)^2}{9} + \frac{(13 - 9)^2}{9} = \frac{16}{9} + \frac{16}{9} = 3.56$$

$$\text{for df} = (3 - 1) = 2 \qquad \chi^2_{(0.10)} = 4.61$$

This value $\chi^2 = 3.56$ is under the 0.10 probability level, so we cannot say there is a relationship between accidents and these months.

EXAMPLE E11-15, *Extended Data on Accidents.* However, if another year gave the same ratios, then we have respectively 26, 18, 10 accidents for these months. The χ^2 value is 7.1, or, there is very positive confidence that a relationship between these months and accidents does exist.

FOR LARGER SETS OF DATA. The definition:

$$\chi^2 = \sum \left[\frac{(O - E)^2}{E} \right]$$

can be applied to larger sets. What we do is to partition the expected value by proportion to the various groups.

Table 11.14   Two Processes with Three Lots of Raw Material (from Ref. 7)*

	Raw Material			
	L	M	N	Totals
Process A	42	13	33	88
Process B	20	8	25	53
Total	62	21	58	141

EXAMPLE E11-16.   Take the series (from Reference 7) as shown in Table 11.14. If the same proportion prevails in all three columns, we have for Process A

$$E \text{ (for cell LA)} = 62 \times 88/141 = 38.7$$
$$E \text{ (for cell MA)} = 21 \times 88/141 = 13.1$$
$$E \text{ (for cell NA)} = 58 \times 88/141 = 36.2$$

Note: $38.7 + 13.1 + 36.2 = 88$

and for Process B

$$E \text{ (for cell LB)} = 62 \times 53/141 = 23.3$$
$$E \text{ (for cell MB)} = 21 \times 53/141 = 7.9$$
$$E \text{ (for cell NB)} = 58 \times 53/141 = 21.8$$

Note: $23.3 + 7.9 + 21.8 = 53.$

Then for chi-square:

$$\chi^2 = \frac{(42 - 38.7)^2}{38.7} + \frac{(13 - 13.1)^2}{13.1} + \frac{(33 - 36.2)^2}{36.2} + \frac{(20 - 23.3)^2}{23.3}$$
$$+ \frac{(8 - 7.9)^2}{7.9} + \frac{(25 - 21.8)^2}{21.8} = 1.5$$

In Table 11.9 the χ^2 value for 0.05 probability is 5.99. Conclusion: There is no significant difference between Process A and Process B.

See Brownlee for a 3×6 test with 10 degrees of freedom $(3 - 1) \times (6 - 1) = 10$.

*From K. A. Brownlee, *Industrial Experimentation*, Chemical Publishing Co., Inc., 1949, by permission of Chemical Publishing Co., Inc. Copyright 1949 by Chemical Publishing Co., Inc.

SUMMARY OF df FOR χ^2 TESTS.   In the chi-square test, the degrees of freedom (df) are given by the number of independent ways the contingency table can be filled up, *given the marginal totals.* In the case below of

Group A	5✓	defective +	120✓	good =	125
Group B	7✓	defective +	43✓	good =	50
Total	12		163	=	175

We can put *any one* of the checked values in the table and then derive all the other checked values by addition or subtraction from the *given marginal totals.* So the system has one df because only one value is required. We put in 5 and get 7, 120, and 43 by difference. This portion of the table (1 df) is of most interest to us.

Another way of getting the degrees of freedom is to take the data without averages of columns or rows, and calculate:

$$(\text{rows} - 1)\ (\text{columns} - 1) = \text{df}$$

For a case we used in Example E11-16, rows = 2; columns = 3, so

$$(2 - 1)(3 - 1) = 2\ \text{df}$$

Other aspects of quality evaluation by attributes are given in Section 12.

12

THE QUALITY EVALUATION OF PRODUCTION LOTS

Quality — a distinctive
and desirable trait often
compromised for quantity.

This section is a preliminary discussion relating to sampling and quality evaluation of large lots. It briefly discusses acceptances and rejections; producer and consumer risk; average quality levels; quality ranges; and leads into some aspects of sampling plans.

Our primary purpose here is not to discuss these subjects in any great detail as this publication is not a quality control manual; the purpose is to present enough of a story so that we lead the reader into a realization of some of the problems to be faced in quality control. These initial concepts will lead to useful views which may influence the early laboratory evaluation and testing phases in which we are mainly interested.

Our earlier chapters have developed a view on the size of sample needed for many *comparisons*. We have assumed that the samples studied were representative of the experiment or process. Now, we look at that "representative" word more carefully. To establish a quality level requires far more samples than to make a comparison.

We must now assume that whatever process we are going to study is under control. This subject is not covered in this manual. What we are going to determine is that chance of sampling wherein the large lot is represented in the sample lot.

One can build up curves of probability (P) versus lot size (n) and rejection limit (c) in several manners. However, when the fraction defective (c/n) is less than 0.05 (5%), and when the sample size is greater than 20 ($n > 20$), we can use the Poisson function for estimation. It is not useful if c/n is zero (perfect quality). The Poisson distribution is of limited use with n as low as

10 and c/n as high as 0.10 (or 10%). Tables of complete summation of terms in this probability function are used to build up the operating characteristic curves for any set of sample conditions.

12.1 CONSTRUCTION OF OPERATING CHARACTERISTIC CURVES

We cannot discuss the Poisson here, but only its usage. For a preliminary discussion, see Reference 28. A complete discussion is in Reference 20 and an industrial example is in Reference 28. The tables of summation of the terms of the Poisson are given in any standard work. (See Ref. 20, pp. 542–546; also Figure 12.1.)

They are tabulated on the basis of an occurrence of c or fewer defectives being found in a sample of n size taken from a large population with p of known quality value (value of fractional defectives is known). The product np is calculated and the chart read.

EXAMPLE E12-1. For instance, take $n = 50$ and $p = 0.02$, then $np = 1.0$. Reading probability versus c from Figure 12.1, we get Table 12.1.

This means that if we should sample and resample our population (throwing the samples back in) 1,000 times, we would have:

368 times when no defect was found
368 times when   1 defect was found
184 times when   2 defects were found
61 times when   3 defects were found
15 times when   4 defects were found
3 times when   5 defects were found

Please note how these figures follow from the normal distribution curve.

Table 12.1   Poisson Series; $p = 0.02, n = 50, np = 100$

c or Fewer Defectives per Lot	Cumulative Probability from Table	Group Probability*
0	0.368	0.368
1	0.736	0.368
2	0.920	0.184
3	0.981	0.061
4	0.996	0.015
5	0.999	0.003
6	1.000	0.001
*Differences		1.000

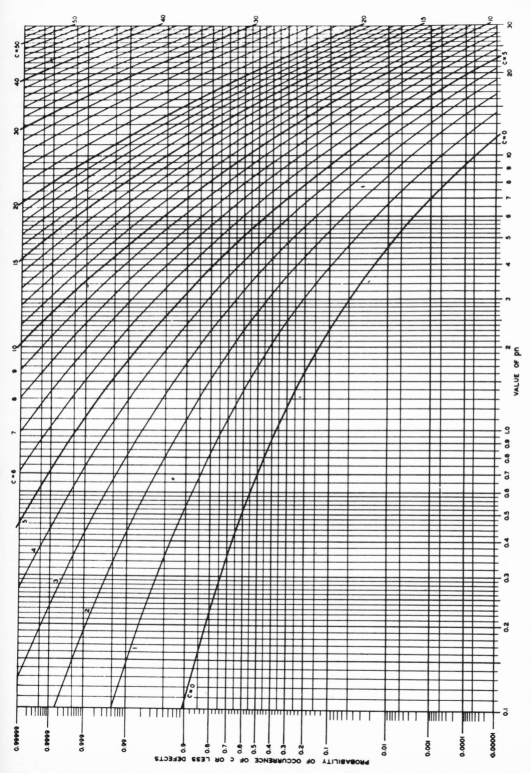

Figure 12.1 Cumulative probability curves from the Poisson exponential. From Reference 14, H. F. Dodge and H. G. Romig, *Sampling Inspection Tables*, Wiley, 1945, by permission of John Wiley & Sons, Inc. Copyright 1945 by John Wiley & Sons, Inc.

Table 12.2 Poisson Cumulative Probability for $n = 50$ and $n = 100$ and Counts from 0 to 3

Fraction Defective		$n = 50$					$n = 100$			
	np	$c = 0$	1	2	3	np	$c = 0$	1	2	3
0.001						0.10	0.905	0.995	1.000	
0.002	0.10	0.905	0.995	1.00		0.20	0.819	0.982	0.999	1.000
0.005	0.25	0.779	0.974	0.998	1.00	0.50	0.607	0.910	0.986	0.998
0.007	0.35	0.705	0.951	0.994	1.00	0.70	0.497	0.844	0.966	0.994
0.01	0.50	0.607	0.910	0.986	0.998	1.00	0.368	0.736	0.920	0.981
0.02	1.0	0.368	0.736	0.920	0.981	2.00	0.135	0.406	0.677	0.857
0.03	1.5	0.223	0.558	0.809	0.934	3.00	0.050	0.199	0.423	0.647
0.04	2.0	0.135	0.406	0.677	0.857	4.00	0.018	0.092	0.238	0.433
0.05						5.00	0.007	0.040	0.125	0.265
0.06	3.0	0.050	0.199	0.423	0.647	6.00	0.002	0.017	0.062	0.151
0.07						7.00	0.001	0.007	0.030	0.082
0.08	4.0	0.018	0.092	0.238	0.433	8.00	0.000	0.003	0.014	0.042
0.09						9.00		0.001	0.006	0.021
0.10	5.0	0.007	0.040	0.125	0.265	10.00			0.003	0.010
0.12	6.0	0.002	0.017	0.062	0.151	12.00			0.001	0.002
0.14	7.0	0.001	0.007	0.030	0.082					
0.16	8.0		0.003	0.014	0.042					
0.18	9.0		0.001	0.006	0.021					
0.20	10.0			0.003	0.010					

We can also read from the Poisson table the probabilities corresponding to $n = 50$ and $c = 1$, and various p values. We calculate np and read the probability for $c = 0$, $c = 1$, etc. See Table 12.2.

Now we shall plot probability against percentage defects in the population as in Figures 12.2, 12.3, and 12.4. These are called *operating characteristic curves* (OCC).

These curves give the probability of acceptance based on n samples and an *acceptance count equal to c*. The entire curve is plotted as a function of the % quality of the grand lot. *Warning:* Sometimes these curves are plotted as a *rejection count (c) equal* to "our c" plus 1.

EXAMPLE E12-2. For instance, from Figure 12.4 we conclude that, using the curve ($n = 110$, $c = 6$), if 7% quality is tested, the sample will have six or fewer defectives 34% of the time. Similarly, this inspection plan will result in accepting batches having 3% defectives 6% of the time.

12.2 INTERPRETATION OF OPERATING CHARACTERISTIC CURVES

Obviously, if we sample a 2% defective lot by taking 50 samples (Figure 12.2) and then counting rejects, and if we have a chance of 368 times in 1000

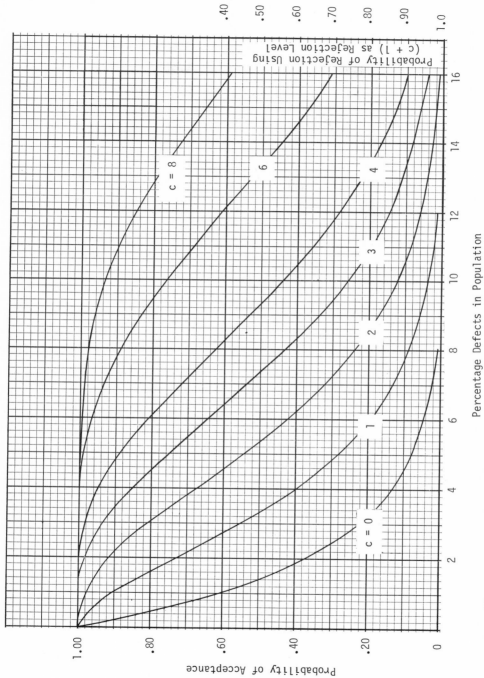

Figure 12.2 Operating characteristic curves, with $n = 50$ and $c = 0$ to 8. See Examples E12-4 and E12-6.

times of getting no defects, we can predict little about the quality. *Sample lots without defects limit quality only very broadly. Quality is not perfect just because no defects were found.*

QUALITY RANGE.   We can, however, predict a quality value where we have a split:

half of such groups of samples are above the level
half of such groups of samples are below the level

This is the point on the OCC where the probability is 0.50. We can also predict the limits of percentage quality which will be accepted, say 90% of the time. For this we read off the percentage defective corresponding to 0.05 and 0.95 probability (5% loss at each end).

EXAMPLE E12-3.   Take the curve for 30 samples (n) and 1 defect accepted (Figure 12.4). We would reject about 1.1% on the lower end ($0.95P$) and also about 5.0% on the upper end ($0.05P$), or a total of 6.1% rejection. The quality would average 5.5% defective. The best quality passed 90% of the time would be 1.0% defective and the worst quality 16% defective. That is, there is some chance of quality better than 1.0% giving a defect in a sample of 30. Likewise, there is some chance of a quality worse than 16% giving no defects in 30 samples.

EXAMPLE E12-4.   Now take the case of 50 samples and $c = 1$ (1 reject accepted in a sample) (Figure 12.2). Here the average quality is 3.4% defective, the low limit (0.95 probability) is 0.7% defective, and the high limit (0.05 probability) is 9.5% defective.

EXAMPLE E12-5.   A case of 100 samples, with $c = 1$, gives (Figure 12.3) approximately

average   1.7%   low limit   0.3%   high limit   4.8%

Thus, we have Table 12.3. Note that the average quality compares to the ratio of c/n as shown in Table 12.4.

Table 12.3   Quality Versus Sample Size with Count = 1 Defective

	$n = 30$ $c = 1$	$n = 50$ $c = 1$	$n = 100$ $c = 1$
Low limit passed	1.0%	0.7%	0.3%
Average quality indicated	5.5%	3.4%	1.7%
High limit passed	16 %	9.5%	4.8%

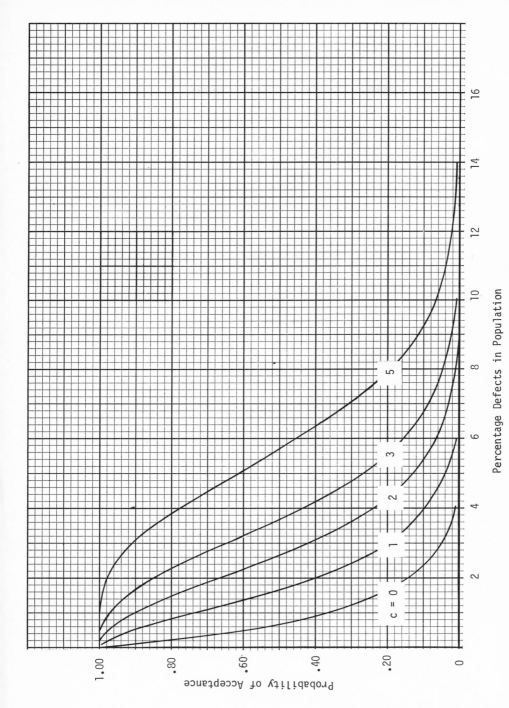

Figure 12.3 Operating characteristic curves, with $n = 100$ and $c = 0$ to 5. See Examples E12-5 and E12-6.

Table 12.4   Sample Size and Average Quality

	$n = 30$	$n = 50$	$n = 100$
Average quality (%)	5.5	3.4	1.7
Ratio c/n (%)	3.3	2.0	1.0

Thus the test tends to pass an average quality considerably worse than the ratio of defectives to sample size.

EXAMPLE E12-6, *Sample Size.*   Let us compare the choice of a sample size of 50 versus 100 units, with the counts of 0 to 8 (Figure 12.2) as in Table 12.5 and the counts of 0 to 5 (Figure 12.3) as in Table 12.6. Note particularly the comparisons of quality value and of range accepted for the sample size of 100 versus 50. The larger sample size has reduced the range greatly. The advantages are evident.

Table 12.5   For Example E12-6 when $n = 50$, $c = 0$ to 8, $pn =$ Variable (from Figure 12.2)

	Quality Value	Limits of Quality		
c	(0.50) % Defective	Lower (0.95) 5% Beyond	Upper (0.05) 5% Beyond	Range Quality Value Accepted
0	1.4	0.1	6.0	0.1 to 6.0 = 5.9
1	3.3	0.7	9.4	0.7 to 9.4 = 8.7
2	5.3	1.5	12.6	1.5 to 12.6 = 11.1
3	7.3	2.7	15.5	2.7 to 15.5 = 12.8
4	9.4	4.0	—	—
6	13.4	6.8	—	—
8	—	9.4	—	—

Table 12.6   For Example E12-6 when $n = 100$, $c = 0$ to 5, $pn =$ Variable (from Figure 12.3)

	Quality Value	Limits of Quality		
c	(0.50) % Defective	Lower (0.95) 5% Beyond	Upper (0.05) 5% Beyond	Range Quality Value Accepted
0	0.7	<0.1	3.0	0.0 to 3.0 = 3.0
1	1.7	0.3	4.8	0.3 to 4.8 = 4.5
2	2.7	0.8	6.4	0.8 to 6.4 = 5.6
3	3.7	1.3	8.0	1.3 to 8.0 = 6.7
5	5.7	2.6	10.6	2.6 to 10.6 = 8.0

Note also that the range from low to high is much narrowed by the larger sample number.

It would be useful to have a table relating these ranges to sample number and defectives. Table 12.7 gives such values based on slightly different confidence of 0.95, and on the binomial instead of the Poisson.

EXAMPLE E12-7. Suppose we have five defectives in 100 samples — we can feel sure that the large lot was between 2% and 11% defective, while 1 in 20 gives us a range of 0% to 25% for the large lot.

We can thus see the advantage of a large sample.

CONSUMER AND PRODUCER RISK. We are not here greatly interested in the consumer-producer problem of quality production and acceptance. But certain factors of this problem lead us to one of considerable interest.

In Figure 12.2 and others for the OCC, we have plotted cumulative probability versus the quality level for certain sample sizes and defective counts. This probability is really a probability for acceptance based on, say, 2 defectives in 50 or any other n and c value. On the opposite edge of Figure 12.2 we have plotted probability for rejection which progresses from 1 to 0 upward. Now the consumer would desire perfect quality while the producer wishes to sell all he makes. But a compromise is used. The consumer usually agrees to accept some percent quality which he can use. He selects a probability P_c, based on some sampling system, such that the quality level passed by this P_c is acceptable to him. The *consumer's risk (CR) is the risk that the consumer takes that, if a lot of some specified quality* (P_t) *is submitted for inspection, the lot will be accepted by his sampling acceptance plan*, when in fact the lot is subquality. We state this as $P_a = 0.10 =$ probability of acceptance is 0.10.

EXAMPLE E12-8. For instance, in Figure 12.4, the consumer's risk is as follows for three curves:

$$n = 110 \quad c = 6 \quad \text{rejectable quality level} = 9\tfrac{1}{2}\%$$
$$n = 500 \quad c = 20 \quad \text{rejectable quality level} = 5\tfrac{1}{2}\%$$
$$n = 50 \quad c = 2 \quad \text{rejectable quality level} = 10\tfrac{1}{2}\%$$

Thus we see that when $CR = P_a = 0.10$ (consumer's risk of acceptance) and $P_a =$ probability of acceptance, then probability of acceptance $= 0.10$. These are the intercepts of probability $= 0.10$ percentage defectives in the population.

This is sometimes called the RQL or rejectable quality level. Thus the OCC tells us that 10% of the time quality of this level will be passed by the sampling plans given above.

How to Use Table 12.7 on Confidence Intervals for Attributes

Table 12.7 gives the general range of relationship between the number of defectives counted with a sample size of n, to the limits of quality which such a combination defines. For instance,

         2 counts in   30 defines quality   1 to 22% (not very close)
         0 counts in   20 defines quality   0 to 17%
         8 counts in   30 defines quality 12 to 46%
        50 counts in   100 defines quality 40 to 60%
        50 counts in   250 defines quality 15 to 26%
       500 counts in 1000 defines quality 47 to 53%

For 50 counts in 250, note that r/n equals 50/250 equals 0.20.

If r exceeds 50, read $100 - r$ = number observed, and subtract each confidence limit from 100.

If r/n exceeds 0.50, read $1.00 - (r/n)$ = fraction observed, and subtract each confidence limit from 100.

Table 12.7 Ninety-five Percent Confidence Intervals for Binomial Distribution (Ref. 35).* Table entries give quality range. See notes on how to use.

No. observed r	Size of sample n 10		15		20		30		50		100		Fraction observed r/n	Size of sample 250		1,000	
0	0	31	0	22	0	17	0	12	0	07	0	4	0.00	0	1	0	0
1	0	45	0	32	0	25	0	17	0	11	0	5	0.01	0	4	0	2
2	3	56	2	40	1	31	1	22	0	14	0	7	0.02	1	5	1	3
3	7	65	4	48	3	38	2	27	1	17	1	8	0.03	1	6	2	4
4	12	74	8	55	6	44	4	31	2	19	1	10	0.04	2	7	3	5
5	19	81	12	62	9	49	6	35	3	22	2	11	0.05	3	9	4	7
6	26	88	16	68	12	54	8	39	5	24	2	12	0.06	3	10	5	8
7	35	93	21	73	15	59	10	43	6	27	3	14	0.07	4	11	6	9
8	44	97	27	79	19	64	12	46	7	29	4	15	0.08	5	12	6	10
9	55	100	32	84	23	68	15	50	9	31	4	16	0.09	6	13	7	11
10	69	100	38	88	27	73	17	53	10	34	5	18	0.10	7	14	8	12
11			45	92	32	77	20	56	12	36	5	19	0.11	7	16	9	13
12			52	96	36	81	23	60	13	38	6	20	0.12	8	17	10	14
13			60	98	41	85	25	63	15	41	7	21	0.13	9	18	11	15
14			68	100	46	88	28	66	16	43	8	22	0.14	10	19	12	16
15			78	100	51	91	31	69	18	44	9	24	0.15	10	20	13	17
16					56	94	34	72	20	46	9	25	0.16	11	21	14	18
17					62	97	37	75	21	48	10	26	0.17	12	22	15	19
18					69	99	40	77	23	50	11	27	0.18	13	23	16	21
19					75	100	44	80	25	53	12	28	0.19	14	24	17	22
20					83	100	47	83	27	55	13	29	0.20	15	26	18	23
21							50	85	28	57	14	30	0.21	16	27	19	24
22							54	88	30	59	14	31	0.22	17	28	19	25
23							57	90	32	61	15	32	0.23	18	29	20	26
24							61	92	34	63	16	33	0.24	19	30	21	27
25							65	94	36	64	17	35	0.25	20	31	22	28
26							69	96	37	66	18	36	0.26	20	32	23	29
27							73	98	39	68	19	37	0.27	21	33	24	30
28							78	99	41	70	19	38	0.28	22	34	25	31
29							83	100	43	72	20	39	0.29	23	35	26	32
30							88	100	45	73	21	40	0.30	24	36	27	33
31									47	75	22	41	0.31	25	37	28	34
32									50	77	23	42	0.32	26	38	29	35
33									52	79	24	43	0.33	27	39	30	36
34									54	80	25	44	0.34	28	40	31	37
35									56	82	26	45	0.35	29	41	32	38
36									57	84	27	46	0.36	30	42	33	39
37									59	85	28	47	0.37	31	43	34	40
38									62	87	28	48	0.38	32	44	35	41
39									64	88	29	49	0.39	33	45	36	42
40									66	90	30	50	0.40	34	46	37	43
41									69	91	31	51	0.41	35	47	38	44
42									71	93	32	52	0.42	36	48	39	45
43									73	94	33	53	0.43	37	49	40	46
44									76	95	34	54	0.44	38	50	41	47
45									78	97	35	55	0.45	39	51	42	48
46									81	98	36	56	0.46	40	52	43	49
47									83	99	37	57	0.47	41	53	44	50
48									86	100	38	58	0.48	42	54	45	51
49									89	100	39	59	0.49	43	55	46	52
50									93	100	40	60	0.50	44	56	47	53

*From E. B. Wilson, *An Introduction to Scientific Research*, McGraw-Hill, 1952, by permission of McGraw-Hill Book Company. Copyright 1952 by McGraw-Hill Book Company.

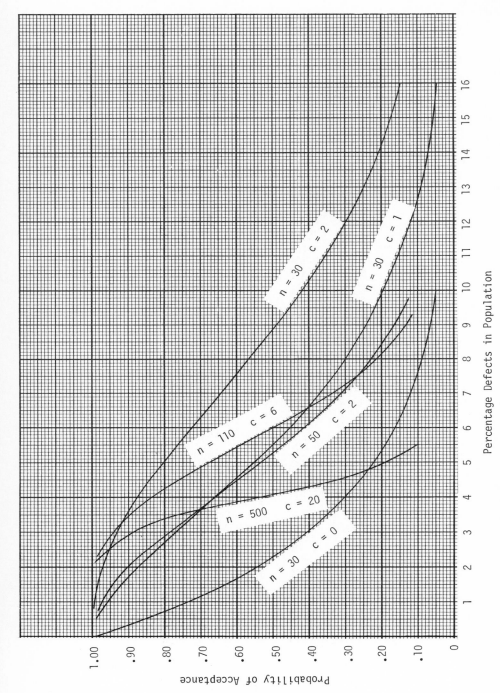

Figure 12.4 Operating characteristic curves when $n = 30$, $c = 0$ to 2; $n = 110$, $c = 6$; $n = 500$, $c = 20$; $n = 50$, $c = 2$. See Examples E12-2, E12-3, and E12-8.

In a similar fashion there is also a *producer's risk*, commonly set at 0.97 or 0.98 P_a. This is at the top end of the operating characteristics curve. The quality level measured at this P_a is called the AQL or *average quality level*. Two or 3% of the time, samples from lots of this quality will result in rejection. Note that the producer's risk results in *rejection of good lots* and the consumer's risk results in *acceptance of poor lots*.

EXAMPLE E12-9.   Thus for the three cases considered above, we have the data given in Table 12.8, which shows that 10% of the time, quality of the rejectable quality level (RQL) will be passed and that 3% of the time, quality of the average quality level (AQL) will be rejected.

Table 12.8   Data for Example E12-9

Sampling Plan		Percent Defective	
		RQL*	AQL**
n	c	0.10P	0.97P
500	20	$5\frac{1}{2}\%$	2.3%
110	6	$9\frac{1}{2}\%$	2.6%
50	2	$10\frac{1}{2}\%$	1.0%

*Consumer's Risk of Acceptance
**Producer's Risk of Rejection

Basically, then, we make comparisons of two probabilities, one near 0.97 and one near 0.10, and find that as these two percentage splits (AQL and RQL) get closer together, we must use very large samples — the curves become steeper.

EXAMPLE E12-10, *Two Interesting Sampling Comparisons*.   Figure 12.5 shows four OCC sampling plans as in Table 12.9. Each plan samples 10% of the large lot, and each plan gives a 0.90 probability of acceptance of a lot which is 1% defective. These curves all meet at about 1% defective. But now you can establish, as we did in Section 12.2, the risks of producer and consumer.

Table 12.9   Effect of Lot Size When Sample Is 10% of Lot

Lot Size N	Sample Size n	c
100	10	0
500	50	1
2,500	250	4
25,000	2,500	31

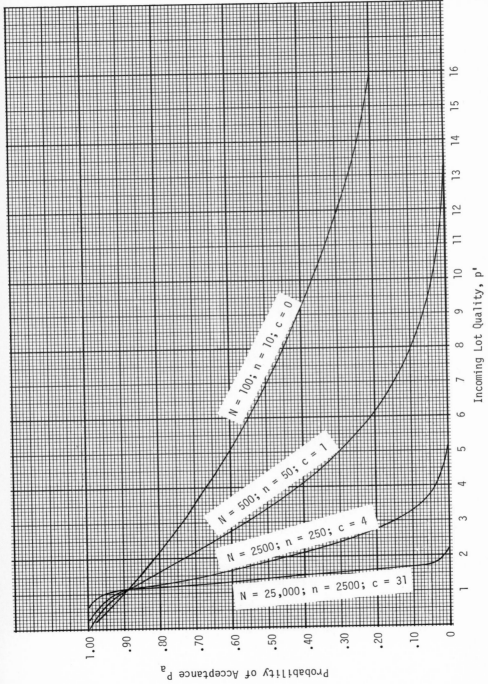

Figure 12.5 Operating characteristic curves for a lot of 10% sampling plans, each of which has a producer's risk of rejection of lots near 1.1% defectives. See Example E12-10.

Incoming Lot Quality, p'

Probability of Acceptance P_a

N = 100; n = 10; c = 0

N = 500; n = 50; c = 1

N = 2500; n = 250; c = 4

N = 25,000; n = 2500; c = 31

EXAMPLE E12-11.   Figure 12.6 shows four OCC sampling plans. See also Table 12.10. In this case the curves all meet at about 5.3% fraction defective, and give a rather narrow range of consumer's risk of acceptance. The range of producer's risk of rejection is broad. Similarly, a plan as shown in Table 12.11 gives a lot tolerance fraction defective (LTFD) of 10% defectives.

Table 12.10   Effect of Sample Size on OCC for Similar Consumer's Risk of Acceptance

N	n	c
3,000	43	0
3,000	102	2
3,000	150	4
3,000	222	7

Table 12.11   Effect of Sample Size on OCC for Similar Producer's Risk of Rejection

N	n	c
1,000	100	0
1,000	170	1
1,000	240	2

EXAMPLE E12-12, *Double Sampling Plans.*   Double sampling plans have been developed to reduce the amount of inspection. A typical one is as given in Table 12.12. This plan is fairly closely reproduced by a single sampling of 100 units, and a count of 1 for rejection. The case is copied from Ref. 25.

Table 12.12   Double Sampling Plan, Example E12-12 (from Ref. 25)

Step	Sample	Action
A	70 units	If 0 defects — *Pass* If 3 or more defects — *Reject* If 1 or 2 defects — *Proceed to Step* B
B	An additional 140 units	If total defectives in both lots are 1 or 2 — *Accept* If 3 or more — *Reject*

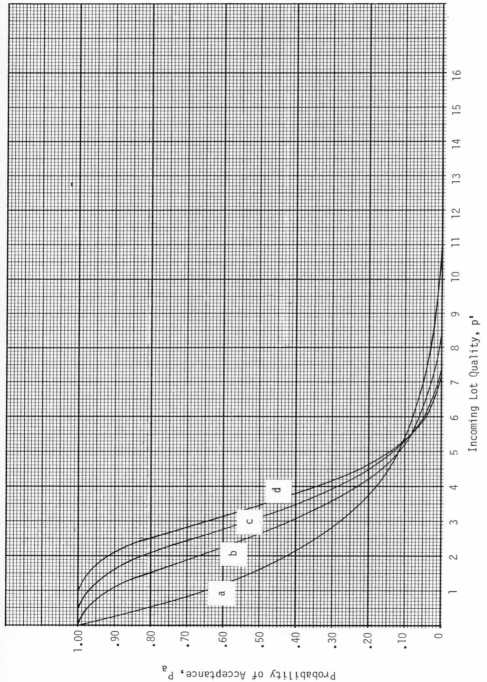

Figure 12.6 Operating characteristic curves for a set of sampling plans each of which has a consumer's risk of acceptance of a lot of 5.3% defectives. When the lot size is 3,000, readings are as follows: curve a, $n = 43$, $c = 0$; curve b, $n = 102$, $c = 2$; curve c, $n = 150$, $c = 4$; curve d, $n = 222$, $c = 7$. See Example E12-11.

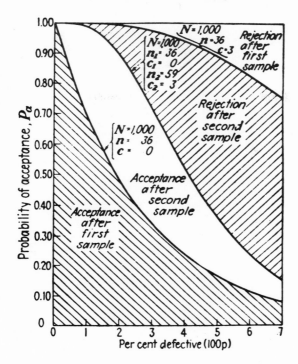

Figure 12.7 Characteristics of a double sampling plan. See Example E12-13. From Reference 20, E. L. Grant, *Statistical Quality Control*, McGraw-Hill, 1946, by permission of McGraw Hill Book Company. Copyright 1946 by McGraw Hill Book Company.

EXAMPLE E12-13. Another plan is illustrated by the OCC in Figure 12.7 (Ref. 20). Here a first sample $n = 36$ is taken. If there are no defectives, the lot of 1,000 is accepted. If there are 4 or more defectives, the lot of 1000 is rejected. If there are counts of 1, 2, or 3 defectives, the resampling proceeds. This resample is a lot of $n = 59$. If the total of defectives for *both lots* is 4 or more, the grand lot of 1,000 is rejected. If the total is 1, 2, or 3, the grand lot is accepted.

The above double sampling plan is approximated by a single lot plan, using 65 samples, and a count of 2 defectives. Thus, if the quality is high and passed on first stage of a double plan, the double sampling plan has reduced the inspection work to almost half (36 samples versus 65 samples).

CHARACTERISTICS OF OCC. The high points of the operating characteristic curves should now be evident.

Steep curves require large sample size.

A double plan can reduce the sampling time.

Always sampling a definite percentage of the lot produces far different results.

Increasing the sample size narrows the confidence range of expected defects.

Curves with definite consumer risk can be made with small samples, but the producer risk then is high, *and vice versa.*

12.3 SEQUENTIAL SAMPLING

A sequential analysis technique allows a minimum of sampling (Ref. 11, 16, and 34). This method plots an upper limit over which rejection occurs, and a lower limit under which acceptance occurs. Between is a range for continued sampling. The common chart shows cumulative number defectives (cumulative c's) plotted versus the cumulative sample number (n).

The two curves are calculated from the desired lot defectives, a probability of acceptance, and a probability of rejection.

EXAMPLE E12-14. Take the case (Ref. 11, p. 339) where

acceptance $= 0.1$ probability of acceptance of a lot with 8% defectives

rejection $= 0.05$ probability of rejection of a lot with 1% defectives

Let us represent the above case as follows:

the producer's risk (α) $= 0.05$
the consumer's risk (β) $= 0.10$
the average number of defectives, or fraction defective,

for producer rejection (m_1) under the α risk $= 0.01$
for consumer acceptance (m_2) under the β risk $= 0.08$

The general equations (Ref. 11) are:

1. The average sample size with lots of quality m_1

$$\text{sample lot size} = \frac{(1 - \alpha)h_1 - \alpha h_2}{s - m_1}$$

2. The average sample size with lots of quality m_2

$$\text{sample lot size} = \frac{(1 - \beta)h_2 - \beta h_1}{m_2 - s}$$

where

$$h_1 = \frac{\ln\,[(1-\alpha)/\beta]}{\ln\,(m_2/m_1)}$$

$$h_2 = \frac{\ln\,[(1-\beta)/\alpha]}{\ln\,(m_2/m_1)}$$

$$s = \frac{m_2 - m_1}{\ln\,(m_2/m_1)}$$

3. The greatest average sample size occurs when the quality m is close to the value of s. It approximates

$$\text{greatest average sample size} = \frac{[h_2 h_1]}{s}$$

To build our diagram of this inspection system, we calculate lines d_1 and d_2 on the greatest and the smallest accumulative numbers of defectives, as follows:

The equation for the lower limit (d_1) line (the line defining immediate acceptance) is

$$d_1 = -[h_1] + sn$$

where

$$h_1 = \frac{\ln\,(0.95/0.1)}{\ln\,(0.08/0.01)} = 1.0826$$

$$s = \frac{(0.08 - 0.01)}{\ln\,(0.08/0.01)} = 0.0337$$

when n = number of items inspected and d_1 = the greatest allowable *cumulative* number of defectives for the lot to be accepted.

The equation for the line for the smallest cumulative number (d_2) of defectives for the lot to be rejected (the line defining immediate rejection) is

$$d_2 = h_2 + sn$$

where

$$h_2 = \frac{\ln\,(0.9/0.05)}{\ln\,(0.08/0.01)} = 1.3900$$

when s = same as above and n = same as above.

The two equations, d_1 and d_2, are linear. We set up Table 12.13 for the values of d_1 and d_2 versus n. The plots for these values are in Figure 12.8. We shall discuss these plots in a moment.

Table 12.13 Sequential Sampling Illustration (from Ref. 11),* based on developed limits

$$d_1 = -1.0826 + 0.0337\,n$$
$$d_2 = 1.3900 + 0.0337\,n$$

n	d_1	d_2
0	-1.08	+1.39
60	+0.94	+3.41
100	+2.29	+4.76

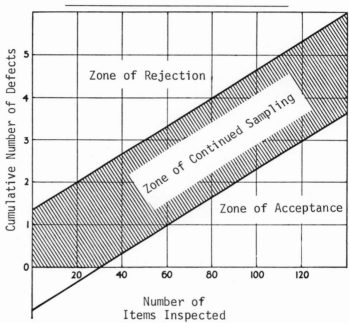

Figure 12.8 Example of sequential sampling based on binomial distribution. Here consumer's risk is $P = 0.10$ for acceptance of a lot with 0.08 (8%) defectives and producer's risk is $P = 0.05$ for rejection of a lot with 0.01 (1%) defectives. Therefore $\alpha = 0.05$, $\beta = 0.1$, $m_1 = 0.01$, $m_2 = 0.08$. See Example E12-14. From Reference 11, O. L. Davies, *Statistical Methods in Research and Production*, Oliver and Boyd, 1961, by permission of Oliver and Boyd on behalf of Imperial Chemical Industries. Copyright 1961, Imperial Chemical Industries.

*From O. L. Davies, *Statistical Methods in Research and Production*, Oliver and Boyd, 1961, by permission of Oliver and Boyd on behalf of Imperial Chemical Industries. Copyright 1961, Imperial Chemical Industries.

The sample size is also estimated from the above criteria for the sampling plan. If the lot has a fraction defective of m_1, the average sample size, using equations above, is

$$\frac{[(1 - \alpha)1.0826] - [\alpha(1.3900)]}{s - m_1}$$

With $m_1 = 0.01$, we get

$$\frac{[(0.95)(1.0826)] - [(.05)(1.3900)]}{0.0337 - 0.01} = \frac{0.97}{0.0237}$$

The sample size is 40 to 41 units. The average sample size with lots of quality m_2 is

$$\frac{(1 - 0.1)1.3900 - (0.1)1.0826}{m_2 - s} = \frac{1.251 - 0.108}{0.08 - 0.0337} = \frac{1.143}{0.046} = 25$$

The greatest average sample size occurs with lots having a quality close to s. From our above discussion, this is

$$\frac{h_1 h_2}{s}$$

For our case, we get

$$\frac{(1.0826)(1.3900)}{0.0337} = \frac{1.5}{0.0337}$$

Greatest average sample size = 44 to 45.

Several statements are positive from the discussion and from Figure 12.8.

The sample size is indicated from the quality, and the consumer's and producer's risk.

Rejection of the lot can only occur when at least two items are inspected — both of which prove defective. This is the intersection of the upper top line with the defective number.

Acceptance cannot be made until at least 32 items are inspected. This is the intersection of the lower line with the number of items inspected.

The procedure is: Sampling is continued as long as the trace of experience falls between the two lines. It is terminated as soon as the trace crosses a line, and appropriate action is indicated.

12.4 THE SAMPLE SIZE (n) FOR QUALITY EVALUATION

We have shown that our immediate interest centers about the low and the high levels of acceptance. We shall use this to develop a concept of sample size for quality evaluation.

Assume that we want a *confidence of rejection* of our lots of 0.90 when quality is inferior. From the Poisson table, we know that it represents for any lot size (n), and acceptance number (c), a definite value of quality. Now as basis for comparison, suppose we are interested in four quality levels, 3%, 5%, 10% and 20% (or 0.03, 0.05, 0.10, 0.20) defectives.

For rough calculations, we shall plot the Poisson function as shown in Figure 12.9. This is convenient for our usage, as we wish to determine values of np, at the same probability (0.90), for different c limits (rejection limits).

How to Use the Chart "Summation of the Poisson" (Figure 12.9)

It is well to view Figure 12.9 as giving the probability of at least c defectives in a sample (do not mention sample size) when the average number of defectives in samples of that size is np. By way of illustration, use the following example (Ref. 33):

What is the least number of samples that one can take from a lot of fraction defective 0.02, so as to have a probability of 0.9 of observing at least 1 defective, and what is the precision of the result?

Quoting from Reference 33*:

Entering Figure 12.9 on the line $P = 0.9$ and finding its intersection with $c = 1$, one would read $np = 2.30$, if one could read the chart that closely. Now $n = 2.30/0.02$. Therefore n is 115. Of course, the probability is 0.9 that the negative error is less than $0.02 - 1/115$ or 0.0113.

One can also readily find the precision of the positive error associated with the specific sample size 115. If the probability is P that one will observe c or more defectives when the average number is np, it follows at once that the probability is $1 - P$ that one will observe $c - 1$ or less defectives under like conditions; therefore, entering Figure 12.9 on the line $P = 1 - 0.9$ or 0.1, and finding its intersection with $pn = 115 \times 0.02$ or 2.30, one finds that this intersection is almost on the line for $c = 5$. One would estimate the value at perhaps 4.8. That is, the probability is 0.1 that 4.8 or more defectives will be observed, because the chart reads that way, viz., at least c. Therefore, the probability is 0.9 that 3.8 or less defectives will be observed under like conditions, viz., sample size 115, $p = 0.02$. In this instance the observed fraction defective would be 3.8 divided by 115 or 0.033. The positive error is $0.033 - 0.02$ or 0.013.

Note: The abscissae in Figure 12.9 is pn or the product of fraction defective and sample number.

Note: By the above procedure, the sample number chart shown in Figure 12.10 was built up.

---

*From L. Simon, *Engineers Manual of Statistical Methods*, Wiley, 1941, by permission of John Wiley & Sons, Inc. Copyright 1941 by John Wiley & Sons, Inc.

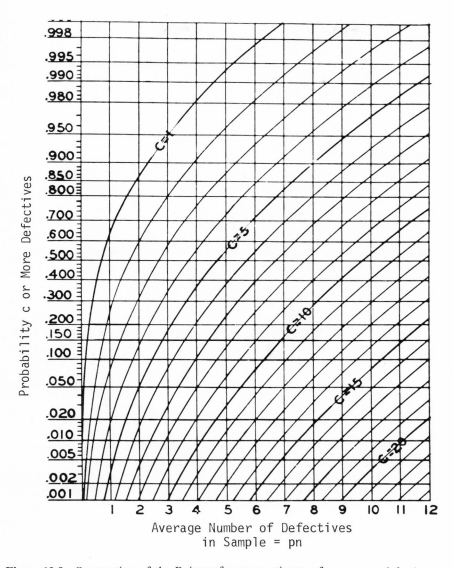

Figure 12.9 Summation of the Poisson for easy estimate of *c* or more defectives. From Reference 33, Simon, L., *Engineers Manual of Statistical Methods*, Wiley, 1941, by permission of John Wiley & Sons, Inc. Copyright 1941 by John Wiley & Sons, Inc.

Table 12.14 Calculation of Sample Size for Different Percent Attribute Quality Levels, where p = fraction defective of grand lot

c Rejection Number	np $P_R = 0.90$ $P_A = 0.10$	Calculate n as follows, when			
		$p = 0.03$	$p = 0.05$	$p = 0.10$	$p = 0.20$
1	2.3	77	46	23	12
2	3.9	130	78	39	20
3	5.3	177	106	53	27
4	6.7	223	134	67	34
5	8.0	267	160	80	40
6	9.2	307	184	92	46
7	10.5	350	210	105	53
8	11.8	393	236	118	59
9	13.0	433	260	130	65
10	14.1	470	282	141	71
12	16.8	560	336	168	84
14	19.0	633	380	190	95

The curve $c = 1$ represents 1 count for rejection, etc. Now we set up a table (Table 12.14). This table becomes useful in arriving at the proper sample size based on quality level of the entire lot.

EXAMPLE E12-15.   It tells us that if we have 3% defectives in our grand lot, we must sample at least 77 units with 9 out of 10 lots giving 1 defective which rejects the lot. In other words, the confidence is that 9 times out of 10 we will reject this 3% defective level or that, if rejection is attained in 9 out of 10 cases, the quality is as expected (0.03).

EXAMPLE E12-16.   Now suppose we reduce our sample to a size of 50 and we get comparable rejections. Then we are confident that our quality is about 5% defectives.

EXAMPLE E12-17.   Now suppose we are going to work to a basis of 5 units rejected. To establish an expectation of

3% defective level — we must sample 267 items
5% defective level — we must sample 160 items
10% defective level — we must sample 80 items
20% defective level — we must sample 40 items

CONCLUSION.   We have a good method to relate sample size to percent defectives. Now we know from Table 12.7 that it takes a fair sample size to confine our percent defectives to a limited range. For a 95% confidence:

0 out of   20 confines % defective to 0 to 17%
1 out of   30 confines % defective to 0 to 17%

3 out of   50 confines % defective to 1 to 17%
7 out of   100 confines % defective to 3 to 14%
20 out of   250 confines % defective to 5 to 12%
90 out of 1000 confines % defective to 7 to 11%

This very type of chart has been calculated. Reference 33 (Chart 0.9) shows how to find the quality level, Q (our p), when the rejection number, c, and the sample number, n, are known. Or if the quality level, Q, is known or assumed, we have a determination of the nearest c line to the Qn (our pn) intersection, on a basis that the sample n will contain fewer than c defectives. The sample size can thus be rather accurately determined. (See Ref. 33, pp. 84 to 93, particularly Figures 10.1, 10.2, 10.3, and chart 0.9.)

It appears that the sample size is usually rather larger than we care to admit. See Table 12.14. For at least 3 defectives, the sample is at least 100 for 5% defective level.

The only alternative is to collect a larger sample. (See following section.)

12.5 RECOMMENDED PROCEDURE FOR DETERMINING SAMPLE SIZE FOR ATTRIBUTES FOR QUALITY EVALUATION

While we can determine significance of differences on smaller samples, our object here is to establish some idea of quality by attributes. To make these illustrations easier to use, the data of Figure 12.1 can be replotted on a basis totally for the estimate of sample size when a certain number of defectives are to be counted. It is basic that if we do not have some defectives, we have no method for estimating quality. The sample size must be so big as to get some defectives. Figure 12.9 is the replot of the Poisson function for this purpose. It plots the *count for rejection* versus pn, or the sample size times the count made.

Continuing our use of 0.90 probability of c or more defectives, we can read off the values, as in Table 12.15, where

$$p = \frac{np \text{ from columns 4 or 5}}{n \text{ from column 3}}$$

Note that columns 1, 2, and 3 are the same as those in Table 12.14. The value (columns 6 and 7) of p carries a tolerance based on 95% confidence. These limits are from Table 8.5 of Reference 35. *These tolerance limits apply in all cases where the observed count is from 1 to 10.* One only needs to recalculate with other p values, and also recalculate column 3 for n.

Table 12.15   Establishing Quality Tolerances

Observed c Count for Rejection	np $(P_r = 0.9)$	If $p = 0.05$, n is	Calculation of Tolerance of p			
			Range of np		Range of p	
			Low	High	Low	High
1	2.3	46	0.0513	4.74	0.0011	0.102
2	3.9	78	0.355	6.30	0.0045	0.081
3	5.3	106	0.818	7.75	0.0077	0.073
4	6.7	134	1.37	9.15	0.0102	0.068
5	8.0	160	1.97	10.51	0.0123	0.066
6	9.2	184	2.61	11.84	0.0142	0.064
7	10.5	210	3.29	13.15	0.0157	0.063
8	11.8	236	3.98	14.43	0.0169	0.061
9	13.0	260	4.70	15.71	0.0181	0.060
10	14.1	282	5.43	16.96	0.0193	0.060

This illustration, of course, comes back to something similar to Table 12.7. Table 12.15 is more specific. *This is really concrete proof* of the value of rather large groups of samples (n) and of a considerable reject number. The range of fraction defectives can be thus limited. However, even with 5 defectives, we have a range of $1\frac{1}{4}\%$ to $6\frac{1}{2}\%$ defective that is possible in the grand lot.

EXAMPLE E12-18.   The only way to limit this further is to use a bigger sample. Table 12.16 presents some typical figures from Reference 33.

Figure 12.10 is a chart allowing prediction of *sample size* based on various probabilities of rejection and on percent quality (1%, 2%, 3%, 5%, 10%, 20% defectives), and plotted as a function of defective counts from 1

How to Use Figure 12.10

Sample size determination, based on attributes, can be made using choice of c counts as basis.

Assume a c count, then travel to right to appropriate P curve, down to appropriate quality level (p value) and left to sample size, n.

EXAMPLE E12-19.   for $c = 3$ and $p = 0.03$

$n = 250$ for $P = 0.98$ and $n = 300$ for $P = 0.995$

EXAMPLE E12-20.   for $c = 5$ and $p = 0.05$

$n = 190$ for $P = 0.95$ and $n = 215$ for $P = 0.98$

Use for sample size greater than 20 ($n > 20$) and fraction defective (p) less than 5%, but not for zero percentage.

Important: This chart is based on a *count for rejection*. See appropriate text.

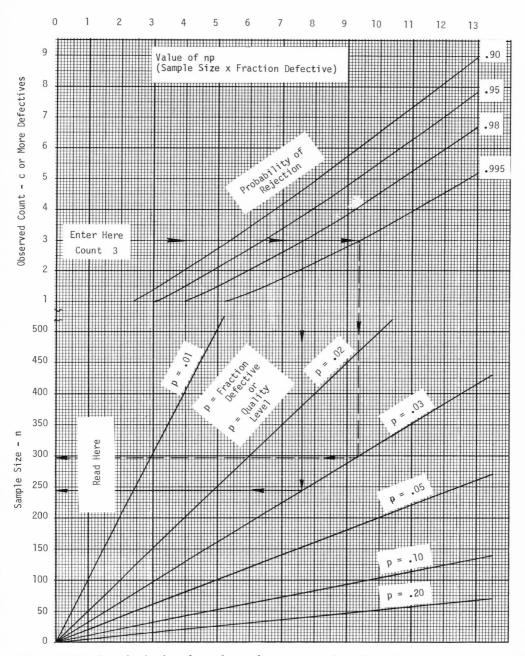

Figure 12.10 Sample size based on observed count, *c*, and as affected by quality level. See Example E12-18.

Table 12.16 Sample (*n*) for Different Percent Defective versus Count of Defectives (from Ref. 33 Chart $I_{.9}$)*

Assume Quality	Using Count Number of Defectives of		
	3	5	7
	Sample Size (*n*) Approximated		
0.20	25	38	50
0.10	51	79	103
0.05	106	160	210
0.03	180	270	350
0.02	265	400	520 (?)
0.01	520 (?)	(?)	(?)

to 9. Note how to use the chart. The illustration enters at 3 count assumed, and leads to a sample size from 240 to 290, with 0.98 to 0.995 probability of rejection and with 3% defectives in lot.

Figure 12.10 is the final result of relating:

quality as *p* or fraction defective at 1%, 2%, 3%, 5%, 10%, 20%
probability of rejection as 0.90, 0.95, 0.98, and 0.995
observed count, *c* or more defectives, from 1 to 9, for rejection
sample size from 0 to 500

Full instructions on the use of Figure 12.10 are given with the figure. Just remember this one thing. These charts are based on a *count* for rejection. Our earlier *c* was based on maximum count for acceptance. Thus

$$c_{\text{for rejection}} = c_{\text{for acceptance}} + 1.$$

12.6   A PRACTICAL EXAMPLE

Recently my young granddaughter ate one of those single serving boxes of cereal which contain raisins. Not a raisin showed up in the box. She told her father, "I don't think they should label this cereal as containing raisins and not put a single raisin into the box." Her father advised her to write the manufacturer. She wrote:

I am a little girl 8 years old. Last week I ate an individual box of your
_____ flakes for breakfast. There was not even one raisin in it. I don't
think you should call it _____ flakes.

The manufacturer replied something like this:

There should have been at least one raisin in the box. In fact, there
should have been several. But no matter how we box this product, there is
always a chance of your getting no raisins.

We are sending you a package of ten boxes for your letter. We hope
they all have raisins in them.

Do you, my reader, want to figure the chances for this case? You might
assume a grand quality level of 10 raisins per box or per 500 flakes of cereal.

REFERENCES

1. Allan, D. H. W., *Statistical Quality Control*. Reinhold, 1959.
2. ASQS Standard A1 (Proposed) Definitions, Symbols, Formulas and Tables for Control Charts. *Ind. Quality Control*, 24(4):217–221, Oct. 1967.
3. A.S.T.M., *Manual on Quality Control of Materials*, Com. E-11, Jan 1951.
4. Barford, N. C., *Experimental Measurements: Precision, Error and Truth*. Addison-Wesley, 1967.
5. Bicking, G. A., Some Use of Statistics in the Planning of Experiments. *Ind. Quality Control*, 10(4):20–24, Jan 1954.
6. Blaedel, W. J., V. W. Meloche, J. A. Ramsay, A Comparison of Criteria for the Rejection of Measurements. *J. Chem. Educ.* 28:643–647, 1951.
7. Brownlee, K. A., *Industrial Experimentation*. Chemical Publishing Co., 1949.
8. Burr, I. W., *Engineering Statistics and Quality Control*. McGraw-Hill, 1953.
9. Close, P., F. C. Raggon, and W. E. Smith. *J. Am. Ceram. Soc.* 33:345, 1950.
10. Cochran, W. A., and G. M. Cox, *Experimental Design*. Wiley, 1950.
11. Davies, O. L., *Statistical Methods in Research and Production*. London: Oliver and Boyd, 1961.
12. Dean, R. B., and W. J. Dixon, Simplified Statistics for Small Numbers of Observations. *Analytical Chem.* 23:636–638, 1951.
13. Dixon, W. J., and F. J. Massey, Jr., *Introduction to Statistical Analysis*, 2d ed. McGraw-Hill, 1957, p. 172.
14. Dodge, H. F., and H. G. Romig, *Sampling Inspection Tables*. Wiley, 1945.
15. Duff, R. D., *J. Am. Ceram. Soc.* 30:12, 1947.
16. Duncan, A. J., *Quality Control and Industrial Statistics*, 3d ed. Irwin, 1965.

17. Eisenhart, C., M. W. Hastay, and W. A. Wallis, *Selected Techniques of Statistical Analysis for Scientific and Industrial Research and Production and Management Engineering*. McGraw-Hill, 1947.

18. Gehring, L. G., *J. Am. Ceram. Soc.* 27:373, 1944.

19. Gore, W. L., *Statistical Methods for Chemical Experimentation*, Interscience, 1952.

20. Grant, E. L., *Statistical Quality Control*. McGraw-Hill, 1946.

21. Hill, L. R. and P. L. Schmidt, Graphical Statistics — An Engineering Approach. *Westinghouse Engr*. March 1950, 120–123; May 1950, 157–160.

22. Hooke, Robert, *Introduction to Scientific Inference*. San Francisco: Holden-Day, 1963.

23. Li, C. H., A Nomograph for Evaluating the Significance of Some Test Results. *A.S.T.M. Bull*. Dec 1953 (Issue 194) T.P. 218, pp. 74–76.

24. Littleton, J. T., *J. Soc. Glass Technol*. 24:180, 1940.

25. Lyon, K. C., Some Aspects of Acceptance Sampling. *Glass Ind*. 33:581, 1952.

26. Natrella, Mary Gibbons, Experimental Statistics. *Natl. Bur. of Std. Handbook*, 91, Aug 1, 1963.

27. Peckner, D., How to Use Probability Paper to Solve Materials Problems. *Mater. in Design Eng*. 52:138, Nov 1960.

28. Preston, F. W., The Laws of Chance: Poisson Series in the Glass Industry and Elsewhere. *Bull. Am. Ceram. Soc*. 28:225–234, 1949.

29. Preston, F. W. *J. Am. Ceram. Soc*. 20:329, 1937.

30. Proschan, F., Experiments Comparing Two Methods for Percentage Defective. *Quality Control Convention Papers*, pp. 43–55, 1954.

31. Richards, J. W., *Interpretation of Technical Data*. Van Nostrand, 1967.

32. Rochester Institute of Technology, *Symbols, Definitions and Tables for Industrial Statistics and Quality Control*. Courtesy, Eastman Kodak Co., rev. 1958.

33. Simon, L., *Engineers Manual of Statistical Methods*. Wiley, 1941.

34. Volk, W. Industrial Statistics. *Chem. Eng*. 63:165, 1956.

35. Wilson, E. B., *An Introduction to Scientific Research*. McGraw-Hill, 1952.

36. Youden, W. J., *Statistical Methods for Chemists*. Wiley, 1951.

APPENDIX

Squares and Square Roots

The tables are reprinted from the A.S.T.M. *Manual on Quality Control of Materials*, Com. E-11 Jan. 1951, pp. 67–73, by permission of the American Society for Testing and Materials. Copyright 1951 by the American Society for Testing and Materials.

To find the square root of numbers greater than 2000, look up the given number in the "Square" column and obtain the answer from the corresponding value in the "No." column.

EXAMPLE 1   To find the square root of 2174.386. Group the digits in pairs starting from the decimal point, thus:

$$\overset{\frown}{21} \quad \overset{\frown}{74}. \quad \overset{\frown}{38} \quad \overset{\frown}{60}$$

There will always be one digit in the square root for each group in the given number. Observe that the square root of the number in the first group (21) is between 4 and 5. Referring to the table, find in the "No." column the two numbers lying between 400 and 500 whose squares most nearly equal the given number. In this case the numbers are 466 and 467. Interpolating and locating the decimal point gives 46.63 for the desired square root.

In referring to the table, numbers between 40 and 50 in the "No." column may be used as well as those between 400 and 500, but the latter numbers will give the desired square root to one more significant figure.

EXAMPLE 2   To find the square root of 21743.86. Group the digits in pairs as before:

$$\overset{\frown}{2} \quad \overset{\frown}{17} \quad \overset{\frown}{43}. \quad \overset{\frown}{86}$$

and observe that the square root of the number in the first group (2) is between 1 and 2. Refer to the "No." column of the table for numbers lying between 1000 and 2000. In this case, the numbers are 1474 and 1475, and interpolating and locating the decimal point gives 147.46.

No.	Square	Square Root	No.	Square	Square Root	No.	Square	Square Root	No.	Square	Square Root
1	1	1.0000	51	2 601	7.1414	101	10 201	10.0499	151	22 801	12.2882
2	4	1.4142	52	2 704	7.2111	102	10 404	10.0995	152	23 104	12.3288
3	9	1.7321	53	2 809	7.2801	103	10 609	10.1489	153	23 409	12.3693
4	16	2.0000	54	2 916	7.3485	104	10 816	10.1980	154	23 716	12.4097
5	25	2.2361	55	3 025	7.4162	105	11 025	10.2470	155	24 025	12.4499
6	36	2.4495	56	3 136	7.4833	106	11 236	10.2956	156	24 336	12.4900
7	49	2.6458	57	3 249	7.5498	107	11 449	10.3441	157	24 649	12.5300
8	64	2.8284	58	3 364	7.6158	108	11 664	10.3923	158	24 964	12.5698
9	81	3.0000	59	3 481	7.6811	109	11 881	10.4403	159	25 281	12.6095
10	100	3.1623	60	3 600	7.7460	110	12 100	10.4881	160	25 600	12.6491
11	121	3.3166	61	3 721	7.8102	111	12 321	10.5357	161	25 921	12.6886
12	144	3.4641	62	3 844	7.8740	112	12 544	10.5830	162	26 244	12.7279
13	169	3.6056	63	3 969	7.9373	113	12 769	10.6301	163	26 569	12.7671
14	196	3.7417	64	4 096	8.0000	114	12 996	10.6771	164	26 896	12.8062
15	225	3.8730	65	4 225	8.0623	115	13 225	10.7238	165	27 225	12.8452
16	256	4.0000	66	4 356	8.1240	116	13 456	10.7703	166	27 556	12.8841
17	289	4.1231	67	4 489	8.1854	117	13 689	10.8167	167	27 889	12.9228
18	324	4.2426	68	4 624	8.2462	118	13 924	10.8628	168	28 224	12.9615
19	361	4.3589	69	4 761	8.3066	119	14 161	10.9087	169	28 561	13.0000
20	400	4.4721	70	4 900	8.3666	120	14 400	10.9545	170	28 900	13.0384
21	441	4.5826	71	5 041	8.4261	121	14 641	11.0000	171	29 241	13.0767
22	484	4.6904	72	5 184	8.4853	122	14 884	11.0454	172	29 584	13.1149
23	529	4.7958	73	5 329	8.5440	123	15 129	11.0905	173	29 929	13.1529
24	576	4.8990	74	5 476	8.6023	124	15 376	11.1355	174	30 276	13.1909
25	625	5.0000	75	5 625	8.6603	125	15 625	11.1803	175	30 625	13.2288
26	676	5.0990	76	5 776	8.7178	126	15 876	11.2250	176	30 976	13.2665
27	729	5.1962	77	5 929	8.7750	127	16 129	11.2694	177	31 329	13.3041
28	784	5.2915	78	6 084	8.8318	128	16 384	11.3137	178	31 684	13.3417
29	841	5.3852	79	6 241	8.8882	129	16 641	11.3578	179	32 041	13.3791
30	900	5.4772	80	6 400	8.9443	130	16 900	11.4018	180	32 400	13.4164
31	961	5.5678	81	6 561	9.0000	131	17 161	11.4455	181	32 761	13.4536
32	1 024	5.6569	82	6 724	9.0554	132	17 424	11.4891	182	33 124	13.4907
33	1 089	5.7446	83	6 889	9.1104	133	17 689	11.5326	183	33 489	13.5277
34	1 156	5.8310	84	7 056	9.1652	134	17 956	11.5758	184	33 856	13.5647
35	1 225	5.9161	85	7 225	9.2195	135	18 225	11.6190	185	34 225	13.6015
36	1 296	6.0000	86	7 396	9.2736	136	18 496	11.6619	186	34 596	13.6382
37	1 369	6.0828	87	7 569	9.3276	137	18 769	11.7047	187	34 969	13.6748
38	1 444	6.1644	88	7 744	9.3808	138	19 044	11.7473	188	35 344	13.7113
39	1 521	6.2450	89	7 921	9.4340	139	19 321	11.7898	189	35 721	13.7477
40	1 600	6.3246	90	8 100	9.4868	140	19 600	11.8322	190	36 100	13.7840
41	1 681	6.4031	91	8 281	9.5394	141	19 881	11.8743	191	36 481	13.8203
42	1 764	6.4807	92	8 464	9.5917	142	20 164	11.9164	192	36 864	13.8564
43	1 849	6.5574	93	8 649	9.6437	143	20 449	11.9583	193	37 249	13.8924
44	1 936	6.6332	94	8 836	9.6954	144	20 736	12.0000	194	37 636	13.9284
45	2 025	6.7082	95	9 025	9.7468	145	21 025	12.0416	195	38 025	13.9642
46	2 116	6.7823	96	9 216	9.7980	146	21 316	12.0830	196	38 416	14.0000
47	2 209	6.8557	97	9 409	9.8489	147	21 609	12.1244	197	38 809	14.0357
48	2 304	6.9282	98	9 604	9.8995	148	21 904	12.1655	198	39 204	14.0712
49	2 401	7.0000	99	9 801	9.9499	149	22 201	12.2066	199	39 601	14.1067
50	2 500	7.0711	100	10 000	10.0000	150	22 500	12.2474	200	40 000	14.1421

No.	Square	Square Root	No.	Square	Square Root	No.	Square	Square Root	No.	Square	Square Root
201	40 401	14.1774	276	76 176	16.6132	351	123 201	18.7350	426	181 476	20.6398
202	40 804	14.2127	277	76 729	16.6433	352	123 904	18.7617	427	182 329	20.6640
203	41 209	14.2478	278	77 284	16.6733	353	124 609	18.7883	428	183 184	20.6882
204	41 616	14.2829	279	77 841	16.7033	354	125 316	18.8149	429	184 041	20.7123
205	42 025	14.3178	280	78 400	16.7332	355	126 025	18.8414	430	184 900	20.7364
206	42 436	14.3527	281	78 961	16.7631	356	126 736	18.8680	431	185 761	20.7605
207	42 849	14.3875	282	79 524	16.7929	357	127 449	18.8944	432	186 624	20.7846
208	43 264	14.4222	283	80 089	16.8226	358	128 164	18.9209	433	187 489	20.8087
209	43 681	14.4568	284	80 656	16.8523	359	128 881	18.9473	434	188 356	20.8327
210	44 100	14.4914	285	81 225	16.8819	360	129 600	18.9737	435	189 225	20.8567
211	44 521	14.5258	286	81 796	16.9115	361	130 321	19.0000	436	190 096	20.8806
212	44 944	14.5602	287	82 369	16.9411	362	131 044	19.0263	437	190 969	20.9045
213	45 369	14.5945	288	82 944	16.9706	363	131 769	19.0526	438	191 844	20.9284
214	45 796	14.6287	289	83 521	17.0000	364	132 496	19.0788	439	192 721	20.9523
215	46 225	14.6629	290	84 100	17.0294	365	133 225	19.1050	440	193 600	20.9762
216	46 656	14.6969	291	84 681	17.0587	366	133 956	19.1311	441	194 481	21.0000
217	47 089	14.7309	292	85 264	17.0880	367	134 689	19.1572	442	195 364	21.0238
218	47 524	14.7648	293	85 849	17.1172	368	135 424	19.1833	443	196 249	21.0476
219	47 961	14.7986	294	86 436	17.1464	369	136 161	19.2094	444	197 136	21.0713
220	48 400	14.8324	295	87 025	17.1756	370	136 900	19.2354	445	198 025	21.0950
221	48 841	14.8661	296	87 616	17.2047	371	137 641	19.2614	446	198 916	21.1187
222	49 284	14.8997	297	88 209	17.2337	372	138 384	19.2873	447	199 809	21.1424
223	49 729	14.9332	298	88 804	17.2627	373	139 129	19.3132	448	200 704	21.1660
224	50 176	14.9666	299	89 401	17.2916	374	139 876	19.3391	449	201 601	21.1896
225	50 625	15.0000	300	90 000	17.3205	375	140 625	19.3649	450	202 500	21.2132
226	51 076	15.0333	301	90 601	17.3494	376	141 376	19.3907	451	203 401	21.2368
227	51 529	15.0665	302	91 204	17.3781	377	142 129	19.4165	452	204 304	21.2603
228	51 984	15.0997	303	91 809	17.4069	378	142 884	19.4422	453	205 209	21.2838
229	52 441	15.1327	304	92 416	17.4356	379	143 641	19.4679	454	206 116	21.3073
230	52 900	15.1658	305	93 025	17.4642	380	144 400	19.4936	455	207 025	21.3307
231	53 361	15.1987	306	93 636	17.4929	381	145 161	19.5192	456	207 936	21.3542
232	53 824	15.2315	307	94 249	17.5214	382	145 924	19.5448	457	208 849	21.3776
233	54 289	15.2643	308	94 864	17.5499	383	146 689	19.5704	458	209 764	21.4009
234	54 756	15.2971	309	95 481	17.5784	384	147 456	19.5959	459	210 681	21.4243
235	55 225	15.3297	310	96 100	17.6068	385	148 225	19.6214	460	211 600	21.4476
236	55 696	15.3623	311	96 721	17.6352	386	148 996	19.6469	461	212 521	21.4709
237	56 169	15.3948	312	97 344	17.6635	387	149 769	19.6723	462	213 444	21.4942
238	56 644	15.4272	313	97 969	17.6918	388	150 544	19.6977	463	214 369	21.5174
239	57 121	15.4596	314	98 596	17.7200	389	151 321	19.7231	464	215 296	21.5407
240	57 600	15.4919	315	99 225	17.7482	390	152 100	19.7484	465	216 225	21.5639
241	58 081	15.5242	316	99 856	17.7764	391	152 881	19.7737	466	217 156	21.5870
242	58 564	15.5563	317	100 489	17.8045	392	153 664	19.7990	467	218 089	21.6102
243	59 049	15.5885	318	101 124	17.8326	393	154 449	19.8242	468	219 024	21.6333
244	59 536	15.6205	319	101 761	17.8606	394	155 236	19.8494	469	219 961	21.6564
245	60 025	15.6525	320	102 400	17.8885	395	156 025	19.8746	470	220 900	21.6795
246	60 516	15.6844	321	103 041	17.9165	396	156 816	19.8997	471	221 841	21.7025
247	61 009	15.7162	322	103 684	17.9444	397	157 609	19.9249	472	222 784	21.7256
248	61 504	15.7480	323	104 329	17.9722	398	158 404	19.9499	473	223 729	21.7486
249	62 001	15.7797	324	104 976	18.0000	399	159 201	19.9750	474	224 676	21.7715
250	62 500	15.8114	325	105 625	18.0278	400	160 000	20.0000	475	225 625	21.7945
251	63 001	15.8430	326	106 276	18.0555	401	160 801	20.0250	476	226 576	21.8174
252	63 504	15.8745	327	106 929	18.0831	402	161 604	20.0499	477	227 529	21.8403
253	64 009	15.9060	328	107 584	18.1108	403	162 409	20.0749	478	228 484	21.8632
254	64 516	15.9374	329	108 241	18.1384	404	163 216	20.0998	479	229 441	21.8861
255	65 025	15.9687	330	108 900	18.1659	405	164 025	20.1246	480	230 400	21.9089
256	65 536	16.0000	331	109 561	18.1934	406	164 836	20.1494	481	231 361	21.9317
257	66 049	16.0312	332	110 224	18.2209	407	165 649	20.1742	482	232 324	21.9545
258	66 564	16.0624	333	110 889	18.2483	408	166 464	20.1990	483	233 289	21.9773
259	67 081	16.0935	334	111 556	18.2757	409	167 281	20.2237	484	234 256	22.0000
260	67 600	16.1245	335	112 225	18.3030	410	168 100	20.2485	485	235 225	22.0227
261	68 121	16.1555	336	112 896	18.3303	411	168 921	20.2731	486	236 196	22.0454
262	68 644	16.1864	337	113 569	18.3576	412	169 744	20.2978	487	237 169	22.0681
263	69 169	16.2173	338	114 244	18.3848	413	170 569	20.3224	488	238 144	22.0907
264	69 696	16.2481	339	114 921	18.4120	414	171 396	20.3470	489	239 121	22.1133
265	70 225	16.2788	340	115 600	18.4391	415	172 225	20.3715	490	240 100	22.1359
266	70 756	16.3095	341	116 281	18.4662	416	173 056	20.3961	491	241 081	22.1585
267	71 289	16.3401	342	116 964	18.4932	417	173 889	20.4206	492	242 064	22.1811
268	71 824	16.3707	343	117 649	18.5203	418	174 724	20.4450	493	243 049	22.2036
269	72 361	16.4012	344	118 336	18.5472	419	175 561	20.4695	494	244 036	22.2261
270	72 900	16.4317	345	119 025	18.5742	420	176 400	20.4939	495	245 025	22.2486
271	73 441	16.4621	346	119 716	18.6011	421	177 241	20.5183	496	246 016	22.2711
272	73 984	16.4924	347	120 409	18.6279	422	178 084	20.5426	497	247 009	22.2935
273	74 529	16.5227	348	121 104	18.6548	423	178 929	20.5670	498	248 004	22 3159
274	75 076	16.5529	349	121 801	18.6815	424	179 776	20.5913	499	249 001	22.3383
275	75 625	16.5831	350	122 500	18.7083	425	180 625	20.6155	500	250 000	22.3607

No.	Square	Square Root	No.	Square	Square Root	No.	Square	Square Root	No.	Square	Square Root
501	251 001	22.3830	576	331 776	24.0000	651	423 801	25.5147	726	527 076	26.9444
502	252 004	22.4054	577	332 929	24.0208	652	425 104	25.5343	727	528 529	26.9629
503	253 009	22.4277	578	334 084	24.0416	653	426 409	25.5539	728	529 984	26.9815
504	254 016	22.4499	579	335 241	24.0624	654	427 716	25.5734	729	531 441	27.0000
505	255 025	22.4722	580	336 400	24.0832	655	429 025	25.5930	730	532 900	27.0185
506	256 036	22.4944	581	337 561	24.1039	656	430 336	25.6125	731	534 361	27.0370
507	257 049	22.5167	582	338 724	24.1247	657	431 649	25.6320	732	535 824	27.0555
508	258 064	22.5389	583	339 889	24.1454	658	432 964	25.6515	733	537 289	27.0740
509	259 081	22.5610	584	341 056	24.1661	659	434 281	25.6710	734	538 756	27.0924
510	260 100	22.5832	585	342 225	24.1868	660	435 600	25.6905	735	540 225	27.1109
511	261 121	22.6053	586	343 396	24.2074	661	436 921	25.7099	736	541 696	27.1293
512	262 144	22.6274	587	344 569	24.2281	662	438 244	25.7294	737	543 169	27.1477
513	263 169	22.6495	588	345 744	24.2487	663	439 569	25.7488	738	544 644	27.1662
514	264 196	22.6716	589	346 921	24.2693	664	440 896	25.7682	739	546 121	27.1846
515	265 225	22.6936	590	348 100	24.2899	665	442 225	25.7876	740	547 600	27.2029
516	266 256	22.7156	591	349 281	24.3105	666	443 556	25.8070	741	549 081	27.2213
517	267 289	22.7376	592	350 464	24.3311	667	444 889	25.8263	742	550 564	27.2397
518	268 324	22.7596	593	351 649	24.3516	668	446 224	25.8457	743	552 049	27.2580
519	269 361	22.7816	594	352 836	24.3721	669	447 561	25.8650	744	553 536	27.2764
520	270 400	22.8035	595	354 025	24.3926	670	448 900	25.8844	745	555 025	27.2947
521	271 441	22.8254	596	355 216	24.4131	671	450 241	25.9037	746	556 516	27.3130
522	272 484	22.8473	597	356 409	24.4336	672	451 584	25.9230	747	558 009	27.3313
523	273 529	22.8692	598	357 604	24.4540	673	452 929	25.9422	748	559 504	27.3496
524	274 576	22.8910	599	358 801	24.4745	674	454 276	25.9615	749	561 001	27.3679
525	275 625	22.9129	600	360 000	24.4949	675	455 625	25.9808	750	562 500	27.3861
526	276 676	22.9347	601	361 201	24.5153	676	456 976	26.0000	751	564 001	27.4044
527	277 729	22.9565	602	362 404	24.5357	677	458 329	26.0192	752	565 504	27.4226
528	278 784	22.9783	603	363 609	24.5561	678	459 684	26.0384	753	567 009	27.4408
529	279 841	23.0000	604	364 816	24.5764	679	461 041	26.0576	754	568 516	27.4591
530	280 900	23.0217	605	366 025	24.5967	680	462 400	26.0768	755	570 025	27.4773
531	281 961	23.0434	606	367 236	24.6171	681	463 761	26.0960	756	571 536	27.4955
532	283 024	23.0651	607	368 449	24.6374	682	465 124	26.1151	757	573 049	27.5136
533	284 089	23.0868	608	369 664	24.6577	683	466 489	26.1343	758	574 564	27.5318
534	285 156	23.1084	609	370 881	24.6779	684	467 856	26.1534	759	576 081	27.5500
535	286 225	23.1301	610	372 100	24.6982	685	469 225	26.1725	760	577 600	27.5681
536	287 296	23.1517	611	373 321	24.7184	686	470 596	26.1916	761	579 121	27.5862
537	288 369	23.1733	612	374 544	24.7386	687	471 969	26.2107	762	580 644	27.6043
538	289 444	23.1948	613	375 769	24.7588	688	473 344	26.2298	763	582 169	27.6225
539	290 521	23.2164	614	376 996	24.7790	689	474 721	26.2488	764	583 696	27.6405
540	291 600	23.2379	615	378 225	24.7992	690	476 100	26.2679	765	585 225	27.6586
541	292 681	23.2594	616	379 456	24.8193	691	477 481	26.2869	766	586 756	27.6767
542	293 764	23.2809	617	380 689	24.8395	692	478 864	26.3059	767	588 289	27.6948
543	294 849	23.3024	618	381 924	24.8596	693	480 249	26.3249	768	589 824	27.7128
544	295 936	23.3238	619	383 161	24.8797	694	481 636	26.3439	769	591 361	27.7308
545	297 025	23.3452	620	384 400	24.8998	695	483 025	26.3629	770	592 900	27.7489
546	298 116	23.3666	621	385 641	24.9199	696	484 416	26.3818	771	594 441	27.7669
547	299 209	23.3880	622	386 884	24.9399	697	485 809	26.4008	772	595 984	27.7849
548	300 304	23.4094	623	388 129	24.9600	698	487 204	26.4197	773	597 529	27.8029
549	301 401	23.4307	624	389 376	24.9800	699	488 601	26.4386	774	599 076	27.8209
550	302 500	23.4521	625	390 625	25.0000	700	490 000	26.4575	775	600 625	27.8388
551	303 601	23.4734	626	391 876	25.0200	701	491 401	26.4764	776	602 176	27.8568
552	304 704	23.4947	627	393 129	25.0400	702	492 804	26.4953	777	603 729	27.8747
553	305 809	23.5160	628	394 384	25.0599	703	494 209	26.5141	778	605 284	27.8927
554	306 916	23.5372	629	395 641	25.0799	704	495 616	26.5330	779	606 841	27.9106
555	308 025	23.5584	630	396 900	25.0998	705	497 025	26.5518	780	608 400	27.9285
556	309 136	23.5797	631	398 161	25.1197	706	498 436	26.5707	781	609 961	27.9464
557	310 249	23.6008	632	399 424	25.1396	707	499 849	26.5895	782	611 524	27.9643
558	311 364	23.6220	633	400 689	25.1595	708	501 264	26.6083	783	613 089	27.9821
559	312 481	23.6432	634	401 956	25.1794	709	502 681	26.6271	784	614 656	28.0000
560	313 600	23.6643	635	403 225	25.1992	710	504 100	26.6458	785	616 225	28.0179
561	314 721	23.6854	636	404 496	25.2190	711	505 521	26.6646	786	617 796	28.0357
562	315 844	23.7065	637	405 769	25.2389	712	506 944	26.6833	787	619 369	28.0535
563	316 969	23.7276	638	407 044	25.2587	713	508 369	26.7021	788	620 944	28.0713
564	318 096	23.7487	639	408 321	25.2784	714	509 796	26.7208	789	622 521	28.0891
565	319 225	23.7697	640	409 600	25.2982	715	511 225	26.7395	790	624 100	28.1069
566	320 356	23.7908	641	410 881	25.3180	716	512 656	26.7582	791	625 681	28.1247
567	321 489	23.8118	642	412 164	25.3377	717	514 089	26.7769	792	627 264	28.1425
568	322 624	23.8328	643	413 449	25.3574	718	515 524	26.7955	793	628 849	28.1603
569	323 761	23.8537	644	414 736	25.3772	719	516 961	26.8142	794	630 436	28.1780
570	324 900	23.8747	645	416 025	25.3969	720	518 400	26.8328	795	632 025	28.1957
571	326 041	23.8956	646	417 316	25.4165	721	519 841	26.8514	796	633 616	28.2135
572	327 184	23.9165	647	418 609	25.4362	722	521 284	26.8701	797	635 209	28.2312
573	328 329	23.9374	648	419 904	25.4558	723	522 729	26.8887	798	636 804	28.2489
574	329 476	23.9583	649	421 201	25.4755	724	524 176	26.9072	799	638 401	28.2666
575	330 625	23.9792	650	422 500	25.4951	725	525 625	26.9258	800	640 000	28.2843

No.	Square	Square Root	No.	Square	Square Root	No.	Square	Square Root	No.	Square	Square Root
801	641 601	28.3019	876	767 376	29.5973	951	904 401	30.8383	1026	1 052 676	32.0312
802	643 204	28.3196	877	769 129	29.6142	952	906 304	30.8545	1027	1 054 729	32.0468
803	644 809	28.3373	878	770 884	29.6311	953	908 209	30.8707	1028	1 056 784	32.0624
804	646 416	28.3549	879	772 641	29.6479	954	910 116	30.8869	1029	1 058 841	32.0780
805	648 025	28.3725	880	774 400	29.6648	955	912 025	30.9031	1030	1 060 900	32.0936
806	649 636	28.3901	881	776 161	29.6816	956	913 936	30.9192	1031	1 062 961	32.1092
807	651 249	28.4077	882	777 924	29.6985	957	915 849	30.9354	1032	1 065 024	32.1248
808	652 864	28.4253	883	779 689	29.7153	958	917 764	30.9516	1033	1 067 089	32.1403
809	654 481	28.4429	884	781 456	29.7321	959	919 681	30.9677	1034	1 069 156	32.1559
810	656 100	28.4605	885	783 225	29.7489	960	921 600	30.9839	1035	1 071 225	32.1714
811	657 721	28.4781	886	784 996	29.7658	961	923 521	31.0000	1036	1 073 296	32.1870
812	659 344	28.4956	887	786 769	29.7825	962	925 444	31.0161	1037	1 075 369	32.2025
813	660 969	28.5132	888	788 544	29.7993	963	927 369	31.0322	1038	1 077 444	32.2180
814	662 596	28.5307	889	790 321	29.8161	964	929 296	31.0483	1039	1 079 521	32.2335
815	664 225	28.5482	890	792 100	29.8329	965	931 225	31.0644	1040	1 081 600	32.2490
816	665 856	28.5657	891	793 881	29.8496	966	933 156	31.0805	1041	1 083 681	32.2645
817	667 489	28.5832	892	795 664	29.8664	967	935 089	31.0966	1042	1 085 764	32.2800
818	669 124	28.6007	893	797 449	29.8831	968	937 024	31.1127	1043	1 087 849	32.2955
819	670 761	28.6182	894	799 236	29.8998	969	938 961	31.1288	1044	1 089 936	32.3110
820	672 400	28.6356	895	801 025	29.9166	970	940 900	31.1448	1045	1 092 025	32.3265
821	674 041	28.6531	896	802 816	29.9333	971	942 841	31.1609	1046	1 094 116	32.3419
822	675 684	28.6705	897	804 609	29.9500	972	944 784	31.1769	1047	1 096 209	32.3574
823	677 329	28.6880	898	806 404	29.9666	973	946 729	31.1929	1048	1 098 304	32.3728
824	678 976	28.7054	899	808 201	29.9833	974	948 676	31.2090	1049	1 100 401	32.3883
825	680 625	28.7228	900	810 000	30.0000	975	950 625	31.2250	1050	1 102 500	32.4037
826	682 276	28.7402	901	811 801	30.0167	976	952 576	31.2410	1051	1 104 601	32.4191
827	683 929	28.7576	902	813 604	30.0333	977	954 529	31.2570	1052	1 106 704	32.4345
828	685 584	28.7750	903	815 409	30.0500	978	956 484	31.2730	1053	1 108 809	32.4500
829	687 241	28.7924	904	817 216	30.0666	979	958 441	31.2890	1054	1 110 916	32.4654
830	688 900	28.8097	905	819 025	30.0832	980	960 400	31.3050	1055	1 113 025	32.4808
831	690 561	28.8271	906	820 836	30.0998	981	962 361	31.3209	1056	1 115 136	32.4962
832	692 224	28.8444	907	822 649	30.1164	982	964 324	31.3369	1057	1 117 249	32.5115
833	693 889	28.8617	908	824 464	30.1330	983	966 289	31.3528	1058	1 119 364	32.5269
834	695 556	28.8791	909	826 281	30.1496	984	968 256	31.3688	1059	1 121 481	32.5423
835	697 225	28.8964	910	828 100	30.1662	985	970 225	31.3847	1060	1 123 600	32.5576
836	698 896	28.9137	911	829 921	30.1828	986	972 196	31.4006	1061	1 125 721	32.5730
837	700 569	28.9310	912	831 744	30.1993	987	974 169	31.4166	1062	1 127 844	32.5883
838	702 244	28.9482	913	833 569	30.2159	988	976 144	31.4325	1063	1 129 969	32.6037
839	703 921	28.9655	914	835 396	30.2324	989	978 121	31.4484	1064	1 132 096	32.6190
840	705 600	28.9828	915	837 225	30.2490	990	980 100	31.4643	1065	1 134 225	32.6343
841	707 281	29.0000	916	839 056	30.2655	991	982 081	31.4802	1066	1 136 356	32.6497
842	708 964	29.0172	917	840 889	30.2820	992	984 064	31.4960	1067	1 138 489	32.6650
843	710 649	29.0345	918	842 724	30.2985	993	986 049	31.5119	1068	1 140 624	32.6803
844	712 336	29.0517	919	844 561	30.3150	994	988 036	31.5278	1069	1 142 761	32.6956
845	714 025	29.0689	920	846 400	30.3315	995	990 025	31.5436	1070	1 144 900	32.7109
846	715 716	29.0861	921	848 241	30.3480	996	992 016	31.5595	1071	1 147 041	32.7261
847	717 409	29.1033	922	850 084	30.3645	997	994 009	31.5753	1072	1 149 184	32.7414
848	719 104	29.1204	923	851 929	30.3809	998	996 004	31.5911	1073	1 151 329	32.7567
849	720 801	29.1376	924	853 776	30.3974	999	998 001	31.6070	1074	1 153 476	32.7719
850	722 500	29.1548	925	855 625	30.4138	1000	1 000 000	31.6228	1075	1 155 625	32.7872
851	724 201	29.1719	926	857 476	30.4302	1001	1 002 001	31.6386	1076	1 157 776	32.8024
852	725 904	29.1890	927	859 329	30.4467	1002	1 004 004	31.6544	1077	1 159 929	32.8177
853	727 609	29.2062	928	861 184	30.4631	1003	1 006 009	31.6702	1078	1 162 084	32.8329
854	729 316	29.2233	929	863 041	30.4795	1004	1 008 016	31.6860	1079	1 164 241	32.8481
855	731 025	29.2404	930	864 900	30.4959	1005	1 010 025	31.7017	1080	1 166 400	32.8634
856	732 736	29.2575	931	866 761	30.5123	1006	1 012 036	31.7175	1081	1 168 561	32.8786
857	734 449	29.2746	932	868 624	30.5287	1007	1 014 049	31.7333	1082	1 170 724	32.8938
858	736 164	29.2916	933	870 489	30.5450	1008	1 016 064	31.7490	1083	1 172 889	32.9090
859	737 881	29.3087	934	872 356	30.5614	1009	1 018 081	31.7648	1084	1 175 056	32.9242
860	739 600	29.3258	935	874 225	30.5778	1010	1 020 100	31.7805	1085	1 177 225	32.9393
861	741 321	29.3428	936	876 096	30.5941	1011	1 022 121	31.7962	1086	1 179 396	32.9545
862	743 044	29.3598	937	877 969	30.6105	1012	1 024 144	31.8119	1087	1 181 569	32.9697
863	744 769	29.3769	938	879 844	30.6268	1013	1 026 169	31.8277	1088	1 183 744	32.9848
864	746 496	29.3939	939	881 721	30.6431	1014	1 028 196	31.8434	1089	1 185 921	33.0000
865	748 225	29.4109	940	883 600	30.6594	1015	1 030 225	31.8591	1090	1 188 100	33.0151
866	749 956	29.4279	941	885 481	30.6757	1016	1 032 256	31.8748	1091	1 190 281	33.0303
867	751 689	29.4449	942	887 364	30.6920	1017	1 034 289	31.8904	1092	1 192 464	33.0454
868	753 424	29.4618	943	889 249	30.7083	1018	1 036 324	31.9061	1093	1 194 649	33.0606
869	755 161	29.4788	944	891 136	30.7246	1019	1 038 361	31.9218	1094	1 196 836	33.0757
870	756 900	29.4958	945	893 025	30.7409	1020	1 040 400	31.9374	1095	1 199 025	33.0908
871	758 641	29.5127	946	894 916	30.7571	1021	1 042 441	31.9531	1096	1 201 216	33.1059
872	760 384	29.5296	947	896 809	30.7734	1022	1 044 484	31.9687	1097	1 203 409	33.1210
873	762 129	29.5466	948	898 704	30.7896	1023	1 046 529	31.9844	1098	1 205 604	33.1361
874	763 876	29.5635	949	900 601	30.8058	1024	1 048 576	32.0000	1099	1 207 801	33.1512
875	765 625	29.5804	950	902 500	30.8221	1025	1 050 625	32.0156	1100	1 210 000	33.1662

No.	Square	Square Root	No.	Square	Square Root	No.	Square	Square Root	No.	Square	Square Root
1101	1 212 201	33.1813	1176	1 382 976	34.2929	1251	1 565 001	35.3695	1326	1 758 276	36.4143
1102	1 214 404	33.1964	1177	1 385 329	34.3074	1252	1 567 504	35.3836	1327	1 760 929	36.4280
1103	1 216 609	33.2114	1178	1 387 684	34.3220	1253	1 570 009	35.3977	1328	1 763 584	36.4417
1104	1 218 816	33.2265	1179	1 390 041	34.3366	1254	1 572 516	35.4119	1329	1 766 241	36.4555
1105	1 221 025	33.2415	1180	1 392 400	34.3511	1255	1 575 025	35.4260	1330	1 768 900	36.4692
1106	1 223 236	33.2566	1181	1 394 761	34.3657	1256	1 577 536	35.4401	1331	1 771 561	36.4829
1107	1 225 449	33.2716	1182	1 397 124	34.3802	1257	1 580 049	35.4542	1332	1 774 224	36.4966
1108	1 227 664	33.2866	1183	1 399 489	34.3948	1258	1 582 564	35.4683	1333	1 776 889	36.5103
1109	1 229 881	33.3017	1184	1 401 856	34.4093	1259	1 585 081	35.4824	1334	1 779 556	36.5240
1110	1 232 100	33.3167	1185	1 404 225	34.4238	1260	1 587 600	35.4965	1335	1 782 225	36.5377
1111	1 234 321	33.3317	1186	1 406 596	34.4384	1261	1 590 121	35.5106	1336	1 784 896	36.5513
1112	1 236 544	33.3467	1187	1 408 969	34.4529	1262	1 592 644	35.5246	1337	1 787 569	36.5650
1113	1 238 769	33.3617	1188	1 411 344	34.4674	1263	1 595 169	35.5387	1338	1 790 244	36.5787
1114	1 240 996	33.3766	1189	1 413 721	34.4819	1264	1 597 696	35.5528	1339	1 792 921	36.5923
1115	1 243 225	33.3916	1190	1 416 100	34.4964	1265	1 600 225	35.5668	1340	1 795 600	36.6060
1116	1 245 456	33.4066	1191	1 418 481	34.5109	1266	1 602 756	35.5809	1341	1 798 281	36.6197
1117	1 247 689	33.4215	1192	1 420 864	34.5254	1267	1 605 289	35.5949	1342	1 800 964	36.6333
1118	1 249 924	33.4365	1193	1 423 249	34.5398	1268	1 607 824	35.6090	1343	1 803 649	36.6470
1119	1 252 161	33.4515	1194	1 425 636	34.5543	1269	1 610 361	35.6230	1344	1 806 336	36.6606
1120	1 254 400	33.4664	1195	1 428 025	34.5688	1270	1 612 900	35.6371	1345	1 809 025	36.6742
1121	1 256 641	33.4813	1196	1 430 416	34.5832	1271	1 615 441	35.6511	1346	1 811 716	36.6879
1122	1 258 884	33.4963	1197	1 432 809	34.5977	1272	1 617 984	35.6651	1347	1 814 409	36.7015
1123	1 261 129	33.5112	1198	1 435 204	34.6121	1273	1 620 529	35.6791	1348	1 817 104	36.7151
1124	1 263 376	33.5261	1199	1 437 601	34.6266	1274	1 623 076	35.6931	1349	1 819 801	36.7287
1125	1 265 625	33.5410	1200	1 440 000	34.6410	1275	1 625 625	35.7071	1350	1 822 500	36.7423
1126	1 267 876	33.5559	1201	1 442 401	34.6554	1276	1 628 176	35.7211	1351	1 825 201	36.7560
1127	1 270 129	33.5708	1202	1 444 804	34.6699	1277	1 630 729	35.7351	1352	1 827 904	36.7696
1128	1 272 384	33.5857	1203	1 447 209	34.6843	1278	1 633 284	35.7491	1353	1 830 609	36.7831
1129	1 274 641	33.6006	1204	1 449 616	34.6987	1279	1 635 841	35.7631	1354	1 833 316	36.7967
1130	1 276 900	33.6155	1205	1 452 025	34.7131	1280	1 638 400	35.7771	1355	1 836 025	36.8103
1131	1 279 161	33.6303	1206	1 454 436	34.7275	1281	1 640 961	35.7911	1356	1 838 736	36.8239
1132	1 281 424	33.6452	1207	1 456 849	34.7419	1282	1 643 524	35.8050	1357	1 841 449	36.8375
1133	1 283 689	33.6601	1208	1 459 264	34.7563	1283	1 646 089	35.8190	1358	1 844 164	36.8511
1134	1 285 956	33.6749	1209	1 461 681	34.7707	1284	1 648 656	35.8329	1359	1 846 881	36.8646
1135	1 288 225	33.6898	1210	1 464 100	34.7851	1285	1 651 225	35.8469	1360	1 849 600	36.8782
1136	1 290 496	33.7046	1211	1 466 521	34.7994	1286	1 653 796	35.8608	1361	1 852 321	36.8917
1137	1 292 769	33.7194	1212	1 468 944	34.8138	1287	1 656 369	35.8748	1362	1 855 044	36.9053
1138	1 295 044	33.7343	1213	1 471 369	34.8281	1288	1 658 944	35.8887	1363	1 857 769	36.9188
1139	1 297 321	33.7491	1214	1 473 796	34.8425	1289	1 661 521	35.9026	1364	1 860 496	36.9324
1140	1 299 600	33.7639	1215	1 476 225	34.8569	1290	1 664 100	35.9166	1365	1 863 225	36.9459
1141	1 301 881	33.7787	1216	1 478 656	34.8712	1291	1 666 681	35.9305	1366	1 865 956	36.9594
1142	1 304 164	33.7935	1217	1 481 089	34.8855	1292	1 669 264	35.9444	1367	1 868 689	36.9730
1143	1 306 449	33.8083	1218	1 483 524	34.8999	1293	1 671 849	35.9583	1368	1 871 424	36.9865
1144	1 308 736	33.8231	1219	1 485 961	34.9142	1294	1 674 436	35.9722	1369	1 874 161	37.0000
1145	1 311 025	33.8378	1220	1 488 400	34.9285	1295	1 677 025	35.9861	1370	1 876 900	37.0135
1146	1 313 316	33.8526	1221	1 490 841	34.9428	1296	1 679 616	36.0000	1371	1 879 641	37.0270
1147	1 315 609	33.8674	1222	1 493 284	34.9571	1297	1 682 209	36.0139	1372	1 882 384	37.0405
1148	1 317 904	33.8821	1223	1 495 729	34.9714	1298	1 684 804	36.0278	1373	1 885 129	37.0540
1149	1 320 201	33.8969	1224	1 498 176	34.9857	1299	1 687 401	36.0416	1374	1 887 876	37.0675
1150	1 322 500	33.9116	1225	1 500 625	35.0000	1300	1 690 000	36.0555	1375	1 890 625	37.0810
1151	1 324 801	33.9264	1226	1 503 076	35.0143	1301	1 692 601	36.0694	1376	1 893 376	37.0945
1152	1 327 104	33.9411	1227	1 505 529	35.0286	1302	1 695 204	36.0832	1377	1 896 129	37.1080
1153	1 329 409	33.9559	1228	1 507 984	35.0428	1303	1 697 809	36.0971	1378	1 898 884	37.1214
1154	1 331 716	33.9706	1229	1 510 441	35.0571	1304	1 700 416	36.1109	1379	1 901 641	37.1349
1155	1 334 025	33.9853	1230	1 512 900	35.0714	1305	1 703 025	36.1248	1380	1 904 400	37.1484
1156	1 336 336	34.0000	1231	1 515 361	35.0856	1306	1 705 636	36.1386	1381	1 907 161	37.1618
1157	1 338 649	34.0147	1232	1 517 824	35.0999	1307	1 708 249	36.1525	1382	1 909 924	37.1753
1158	1 340 964	34.0294	1233	1 520 289	35.1141	1308	1 710 864	36.1663	1383	1 912 689	37.1887
1159	1 343 281	34.0441	1234	1 522 756	35.1283	1309	1 713 481	36.1801	1384	1 915 456	37.2022
1160	1 345 600	34.0588	1235	1 525 225	35.1426	1310	1 716 100	36.1939	1385	1 918 225	37.2156
1161	1 347 921	34.0735	1236	1 527 696	35.1568	1311	1 718 721	36.2077	1386	1 920 996	37.2290
1162	1 350 244	34.0881	1237	1 530 169	35.1710	1312	1 721 344	36.2215	1387	1 923 769	37.2424
1163	1 352 569	34.1028	1238	1 532 644	35.1852	1313	1 723 969	36.2353	1388	1 926 544	37.2559
1164	1 354 896	34.1174	1239	1 535 121	35.1994	1314	1 726 596	36.2491	1389	1 929 321	37.2693
1165	1 357 225	34.1321	1240	1 537 600	35.2136	1315	1 729 225	36.2629	1390	1 932 100	37.2827
1166	1 359 556	34.1467	1241	1 540 081	35.2278	1316	1 731 856	36.2767	1391	1 934 881	37.2961
1167	1 361 889	34.1614	1242	1 542 564	35.2420	1317	1 734 489	36.2905	1392	1 937 664	37.3095
1168	1 364 224	34.1760	1243	1 545 049	35.2562	1318	1 737 124	36.3043	1393	1 940 449	37.3229
1169	1 366 561	34.1906	1244	1 547 536	35.2704	1319	1 739 761	36.3180	1394	1 943 236	37.3363
1170	1 368 900	34.2053	1245	1 550 025	35.2846	1320	1 742 400	36.3318	1395	1 946 025	37.3497
1171	1 371 241	34.2199	1246	1 552 516	35.2987	1321	1 745 041	36.3456	1396	1 948 816	37.3631
1172	1 373 584	34.2345	1247	1 555 009	35.3129	1322	1 747 684	36.3593	1397	1 951 609	37.3765
1173	1 375 929	34.2491	1248	1 557 504	35.3270	1323	1 750 329	36.3731	1398	1 954 404	37.3898
1174	1 378 276	34.2637	1249	1 560 001	35.3412	1324	1 752 976	36.3868	1399	1 957 201	37.4032
1175	1 380 625	34.2783	1250	1 562 500	35.3553	1325	1 755 625	36.4005	1400	1 960 000	37.4166

No.	Square	Square Root	No.	Square	Square Root	No.	Square	Square Root	No.	Square	Square Root
1401	1 962 801	37.4299	1476	2 178 576	38.4187	1551	2 405 601	39.3827	1626	2 643 876	40.3237
1402	1 965 604	37.4433	1477	2 181 529	38.4318	1552	2 408 704	39.3954	1627	2 647 129	40.3361
1403	1 968 409	37.4566	1478	2 184 484	38.4448	1553	2 411 809	39.4081	1628	2 650 384	40.3485
1404	1 971 216	37.4700	1479	2 187 441	38.4578	1554	2 414 916	39.4208	1629	2 653 641	40.3609
1405	1 974 025	37.4833	1480	2 190 400	38.4708	1555	2 418 025	39.4335	1630	2 656 900	40.3733
1406	1 976 836	37.4967	1481	2 193 361	38.4838	1556	2 421 136	39.4462	1631	2 660 161	40.3856
1407	1 979 649	37.5100	1482	2 196 324	38.4968	1557	2 424 249	39.4588	1632	2 663 424	40.3980
1408	1 982 464	37.5233	1483	2 199 289	38.5097	1558	2 427 364	39.4715	1633	2 666 689	40.4104
1409	1 985 281	37.5366	1484	2 202 256	38.5227	1559	2 430 481	39.4842	1634	2 669 956	40.4228
1410	1 988 100	37.5500	1485	2 205 225	38.5357	1560	2 433 600	39.4968	1635	2 673 225	40.4351
1411	1 990 921	37.5633	1486	2 208 196	38.5487	1561	2 436 721	39.5095	1636	2 676 496	40.4475
1412	1 993 744	37.5766	1487	2 211 169	38.5616	1562	2 439 844	39.5221	1637	2 679 769	40.4599
1413	1 996 569	37.5899	1488	2 214 144	38.5746	1563	2 442 969	39.5348	1638	2 683 044	40.4722
1414	1 999 396	37.6032	1489	2 217 121	38.5876	1564	2 446 096	39.5474	1639	2 686 321	40.4846
1415	2 002 225	37.6165	1490	2 220 100	38.6005	1565	2 449 225	39.5601	1640	2 689 600	40.4969
1416	2 005 056	37.6298	1491	2 223 081	38.6135	1566	2 452 356	39.5727	1641	2 692 881	40.5093
1417	2 007 889	37.6431	1492	2 226 064	38.6264	1567	2 455 489	39.5854	1642	2 696 164	40.5216
1418	2 010 724	37.6563	1493	2 229 049	38.6394	1568	2 458 624	39.5980	1643	2 699 449	40.5339
1419	2 013 561	37.6696	1494	2 232 036	38.6523	1569	2 461 761	39.6106	1644	2 702 736	40.5463
1420	2 016 400	37.6829	1495	2 235 025	38.6652	1570	2 464 900	39.6232	1645	2 706 025	40.5586
1421	2 019 241	37.6962	1496	2 238 016	38.6782	1571	2 468 041	39.6358	1646	2 709 316	40.5709
1422	2 022 084	37.7094	1497	2 241 009	38.6911	1572	2 471 184	39.6485	1647	2 712 609	40.5832
1423	2 024 929	37.7227	1498	2 244 004	38.7040	1573	2 474 329	39.6611	1648	2 715 904	40.5956
1424	2 027 776	37.7359	1499	2 247 001	38.7169	1574	2 477 476	39.6737	1649	2 719 201	40.6079
1425	2 030 625	37.7492	1500	2 250 000	38.7298	1575	2 480 625	39.6863	1650	2 722 500	40.6202
1426	2 033 476	37.7624	1501	2 253 001	38.7427	1576	2 483 776	39.6989	1651	2 725 801	40.6325
1427	2 036 329	37.7757	1502	2 256 004	38.7556	1577	2 486 929	39.7115	1652	2 729 104	40.6448
1428	2 039 184	37.7889	1503	2 259 009	38.7685	1578	2 490 084	39.7240	1653	2 732 409	40.6571
1429	2 042 041	37.8021	1504	2 262 016	38.7814	1579	2 493 241	39.7366	1654	2 735 716	40.6694
1430	2 044 900	37.8153	1505	2 265 025	38.7943	1580	2 496 400	39.7492	1655	2 739 025	40.6817
1431	2 047 761	37.8286	1506	2 268 036	38.8072	1581	2 499 561	39.7618	1656	2 742 336	40.6940
1432	2 050 624	37.8418	1507	2 271 049	38.8201	1582	2 502 724	39.7744	1657	2 745 649	40.7063
1433	2 053 489	37.8550	1508	2 274 064	38.8330	1583	2 505 889	39.7869	1658	2 748 964	40.7185
1434	2 056 356	37.8682	1509	2 277 081	38.8458	1584	2 509 056	39.7995	1659	2 752 281	40.7308
1435	2 059 225	37.8814	1510	2 280 100	38.8587	1585	2 512 225	39.8121	1660	2 755 600	40.7431
1436	2 062 096	37.8946	1511	2 283 121	38.8716	1586	2 515 396	39.8246	1661	2 758 921	40.7554
1437	2 064 969	37.9078	1512	2 286 144	38.8844	1587	2 518 569	39.8372	1662	2 762 244	40.7676
1438	2 067 844	37.9210	1513	2 289 169	38.8973	1588	2 521 744	39.8497	1663	2 765 569	40.7799
1439	2 070 721	37.9342	1514	2 292 196	38.9102	1589	2 524 921	39.8623	1664	2 768 896	40.7922
1440	2 073 600	37.9473	1515	2 295 225	38.9230	1590	2 528 100	39.8748	1665	2 772 225	40.8044
1441	2 076 481	37.9605	1516	2 298 256	38.9358	1591	2 531 281	39.8873	1666	2 775 556	40.8167
1442	2 079 364	37.9737	1517	2 301 289	38.9487	1592	2 534 464	39.8999	1667	2 778 889	40.8289
1443	2 082 249	37.9868	1518	2 304 324	38.9615	1593	2 537 649	39.9124	1668	2 782 224	40.8412
1444	2 085 136	38.0000	1519	2 307 361	38.9744	1594	2 540 836	39.9249	1669	2 785 561	40.8534
1445	2 088 025	38.0132	1520	2 310 400	38.9872	1595	2 544 025	39.9375	1670	2 788 900	40.8656
1446	2 090 916	38.0263	1521	2 313 441	39.0000	1596	2 547 216	39.9500	1671	2 792 241	40.8779
1447	2 093 809	38.0395	1522	2 316 484	39.0128	1597	2 550 409	39.9625	1672	2 795 584	40.8901
1448	2 096 704	38.0526	1523	2 319 529	39.0256	1598	2 553 604	39.9750	1673	2 798 929	40.9023
1449	2 099 601	38.0657	1524	2 322 576	39.0384	1599	2 556 801	39.9875	1674	2 802 276	40.9145
1450	2 102 500	38.0789	1525	2 325 625	39.0512	1600	2 560 000	40.0000	1675	2 805 625	40.9268
1451	2 105 401	38.0920	1526	2 328 676	39.0640	1601	2 563 201	40.0125	1676	2 808 976	40.9390
1452	2 108 304	38.1051	1527	2 331 729	39.0768	1602	2 566 404	40.0250	1677	2 812 329	40.9512
1453	2 111 209	38.1182	1528	2 334 784	39.0896	1603	2 569 609	40.0375	1678	2 815 684	40.9634
1454	2 114 116	38.1314	1529	2 337 841	39.1024	1604	2 572 816	40.0500	1679	2 819 041	40.9756
1455	2 117 025	38.1445	1530	2 340 900	39.1152	1605	2 576 025	40.0625	1680	2 822 400	40.9878
1456	2 119 936	38.1576	1531	2 343 961	39.1280	1606	2 579 236	40.0749	1681	2 825 761	41.0000
1457	2 122 849	38.1707	1532	2 347 024	39.1408	1607	2 582 449	40.0874	1682	2 829 124	41.0122
1458	2 125 764	38.1838	1533	2 350 089	39.1535	1608	2 585 664	40.0999	1683	2 832 489	41.0244
1459	2 128 681	38.1969	1534	2 353 156	39.1663	1609	2 588 881	40.1123	1684	2 835 856	41.0366
1460	2 131 600	38.2099	1535	2 356 225	39.1791	1610	2 592 100	40.1248	1685	2 839 225	41.0488
1461	2 134 521	38.2230	1536	2 359 296	39.1918	1611	2 595 321	40.1373	1686	2 842 596	41.0609
1462	2 137 444	38.2361	1537	2 362 369	39.2046	1612	2 598 544	40.1497	1687	2 845 969	41.0731
1463	2 140 369	38.2492	1538	2 365 444	39.2173	1613	2 601 769	40.1622	1688	2 849 344	41.0853
1464	2 143 296	38.2623	1539	2 368 521	39.2301	1614	2 604 996	40.1746	1689	2 852 721	41.0974
1465	2 146 225	38.2753	1540	2 371 600	39.2428	1615	2 608 225	40.1871	1690	2 856 100	41.1096
1466	2 149 156	38.2884	1541	2 374 681	39.2556	1616	2 611 456	40.1995	1691	2 859 481	41.1218
1467	2 152 089	38.3014	1542	2 377 764	39.2683	1617	2 614 689	40.2119	1692	2 862 864	41.1339
1468	2 155 024	38.3145	1543	2 380 849	39.2810	1618	2 617 924	40.2244	1693	2 866 249	41.1461
1469	2 157 961	38.3275	1544	2 383 936	39.2938	1619	2 621 161	40.2368	1694	2 869 636	41.1582
1470	2 160 900	38.3406	1545	2 387 025	39.3065	1620	2 624 400	40.2492	1695	2 873 025	41.1704
1471	2 163 841	38.3536	1546	2 390 116	39.3192	1621	2 627 641	40.2616	1696	2 876 416	41.1825
1472	2 166 784	38.3667	1547	2 393 209	39.3319	1622	2 630 884	40.2741	1697	2 879 809	41.1947
1473	2 169 729	38.3797	1548	2 396 304	39.3446	1623	2 634 129	40.2865	1698	2 883 204	41.2068
1474	2 172 676	38.3927	1549	2 399 401	39.3573	1624	2 637 376	40.2989	1699	2 886 601	41.2189
1475	2 175 625	38.4057	1550	2 402 500	39.3700	1625	2 640 625	40.3113	1700	2 890 000	41.2311

No.	Square	Square Root	No.	Square	Square Root	No.	Square	Square Root	No.	Square	Square Root
1701	2 893 401	41.2432	1776	3 154 176	42.1426	1851	3 426 201	43.0232	1926	3 709 476	43.8862
1702	2 896 804	41.2553	1777	3 157 729	42.1545	1852	3 429 904	43.0349	1927	3 713 329	43.8976
1703	2 900 209	41.2674	1778	3 161 284	42.1663	1853	3 433 609	43.0465	1928	3 717 184	43.9090
1704	2 903 616	41.2795	1779	3 164 841	42.1782	1854	3 437 316	43.0581	1929	3 721 041	43.9204
1705	2 907 025	41.2916	1780	3 168 400	42.1900	1855	3 441 025	43.0697	1930	3 724 900	43.9318
1706	2 910 436	41.3038	1781	3 171 961	42.2019	1856	3 444 736	43.0813	1931	3 728 761	43.9431
1707	2 913 849	41.3159	1782	3 175 524	42.2137	1857	3 448 449	43.0929	1932	3 732 624	43.9545
1708	2 917 264	41.3280	1783	3 179 089	42.2256	1858	3 452 164	43.1045	1933	3 736 489	43.9659
1709	2 920 681	41.3401	1784	3 182 656	42.2374	1859	3 455 881	43.1161	1934	3 740 356	43.9773
1710	2 924 100	41.3521	1785	3 186 225	42.2493	1860	3 459 600	43.1277	1935	3 744 225	43.9886
1711	2 927 521	41.3642	1786	3 189 796	42.2611	1861	3 463 321	43.1393	1936	3 748 096	44.0000
1712	2 930 944	41.3763	1787	3 193 369	42.2729	1862	3 467 044	43.1509	1937	3 751 969	44.0114
1713	2 934 369	41.3884	1788	3 196 944	42.2847	1863	3 470 769	43.1625	1938	3 755 844	44.0227
1714	2 937 796	41.4005	1789	3 200 521	42.2966	1864	3 474 496	43.1741	1939	3 759 721	44.0341
1715	2 941 225	41.4126	1790	3 204 100	42.3084	1865	3 478 225	43.1856	1940	3 763 600	44.0454
1716	2 944 656	41.4246	1791	3 207 681	42.3202	1866	3 481 956	43.1972	1941	3 767 481	44.0568
1717	2 948 089	41.4367	1792	3 211 264	42.3320	1867	3 485 689	43.2088	1942	3 771 364	44.0681
1718	2 951 524	41.4488	1793	3 214 849	42.3438	1868	3 489 424	43.2204	1943	3 775 249	44.0795
1719	2 954 961	41.4608	1794	3 218 436	42.3556	1869	3 493 161	43.2319	1944	3 779 136	44.0908
1720	2 958 400	41.4729	1795	3 222 025	42.3674	1870	3 496 900	43.2435	1945	3 783 025	44 1022
1721	2 961 841	41.4849	1796	3 225 616	42.3792	1871	3 500 641	43.2551	1946	3 786 916	44.1135
1722	2 965 284	41.4970	1797	3 229 209	42.3910	1872	3 504 384	43.2666	1947	3 790 809	44.1248
1723	2 968 729	41.5090	1798	3 232 804	42.4028	1873	3 508 129	43.2782	1948	3 794 704	44.1362
1724	2 972 176	41.5211	1799	3 236 401	42.4146	1874	3 511 876	43.2897	1949	3 798 601	44.1475
1725	2 975 625	41.5331	1800	3 240 000	42.4264	1875	3 515 625	43.3013	1950	3 802 500	44.1588
1726	2 979 076	41.5452	1801	3 243 601	42.4382	1876	3 519 376	43.3128	1951	3 806 401	44.1701
1727	2 982 529	41.5572	1802	3 247 204	42.4500	1877	3 523 129	43.3244	1952	3 810 304	44.1814
1728	2 985 984	41.5692	1803	3 250 809	42.4617	1878	3 526 884	43.3359	1953	3 814 209	44.1928
1729	2 989 441	41.5812	1804	3 254 416	42.4735	1879	3 530 641	43.3474	1954	3 818 116	44.2041
1730	2 992 900	41.5933	1805	3 258 025	42.4853	1880	3 534 400	43.3590	1955	3 822 025	44.2154
1731	2 996 361	41 6053	1806	3 261 636	42.4971	1881	3 538 161	43.3705	1956	3 825 936	44.2267
1732	2 999 824	41.6173	1807	3 265 249	42.5088	1882	3 541 924	43.3820	1957	3 829 849	44.2380
1733	3 003 289	41 6293	1808	3 268 864	42.5206	1883	3 545 689	43.3935	1958	3 833 764	44.2493
1734	3 006 756	41.6413	1809	3 272 481	42.5323	1884	3 549 456	43.4051	1959	3 837 681	44.2606
1735	3 010 225	41.6533	1810	3 276 100	42.5441	1885	3 553 225	43.4166	1960	3 841 600	44.2719
1736	3 013 696	41.6653	1811	3 279 721	42.5558	1886	3 556 996	43.4281	1961	3 845 521	44.2832
1737	3 017 169	41.6773	1812	3 283 344	42.5676	1887	3 560 769	43.4396	1962	3 849 444	44.2945
1738	3 020 644	41.6893	1813	3 286 969	42.5793	1888	3 564 544	43.4511	1963	3 853 369	44.3058
1739	3 024 121	41.7013	1814	3 290 596	42.5911	1889	3 568 321	43.4626	1964	3 857 296	44.3170
1740	3 027 600	41.7133	1815	3 294 225	42.6028	1890	3 572 100	43.4741	1965	3 861 225	44.3283
1741	3 031 081	41.7253	1816	3 297 856	42.6146	1891	3 575 881	43.4856	1966	3 865 156	44.3396
1742	3 034 564	41.7373	1817	3 301 489	42.6263	1892	3 579 664	43.4971	1967	3 869 089	44.3509
1743	3 038 049	41.7493	1818	3 305 124	42.6380	1893	3 583 449	43.5086	1968	3 873 024	44.3621
1744	3 041 536	41.7612	1819	3 308 761	42.6497	1894	3 587 236	43.5201	1969	3 876 961	44.3734
1745	3 045 025	41.7732	1820	3 312 400	42.6615	1895	3 591 025	43.5316	1970	3 880 900	44.3847
1746	3 048 516	41.7852	1821	3 316 041	42.6732	1896	3 594 816	43.5431	1971	3 884 841	44.3959
1747	3 052 009	41.7971	1822	3 319 684	42.6849	1897	3 598 609	43.5546	1972	3 888 784	44.4072
1748	3 055 504	41.8091	1823	3 323 329	42.6966	1898	3 602 404	43.5660	1973	3 892 729	44.4185
1749	3 059 001	41.8210	1824	3 326 976	42.7083	1899	3 606 201	43.5775	1974	3 896 676	44.4297
1750	3 062 500	41.8330	1825	3 330 625	42.7200	1900	3 610 000	43.5890	1975	3 900 625	44.4410
1751	3 066 001	41.8450	1826	3 334 276	42.7317	1901	3 613 801	43.6005	1976	3 904 576	44.4522
1752	3 069 504	41.8569	1827	3 337 929	42.7434	1902	3 617 604	43.6119	1977	3 908 529	44.4635
1753	3 073 009	41 8688	1828	3 341 584	42.7551	1903	3 621 409	43.6234	1978	3 912 484	44.4747
1754	3 076 516	41.8808	1829	3 345 241	42.7668	1904	3 625 216	43.6348	1979	3 916 441	44.4860
1755	3 080 025	41.8927	1830	3 348 900	42.7785	1905	3 629 025	43.6463	1980	3 920 400	44.4972
1756	3 083 536	41.9047	1831	3 352 561	42.7902	1906	3 632 836	43.6578	1981	3 924 361	44.5084
1757	3 087 049	41.9166	1832	3 356 224	42.8019	1907	3 636 649	43.6692	1982	3 928 324	44.5197
1758	3 090 564	41.9285	1833	3 359 889	42.8135	1908	3 640 464	43.6807	1983	3 932 289	44.5309
1759	3 094 081	41.9404	1834	3 363 556	42.8252	1909	3 644 281	43.6921	1984	3 936 256	44.5421
1760	3 097 600	41.9524	1835	3 367 225	42.8369	1910	3 648 100	43.7035	1985	3 940 225	44.5533
1761	3 101 121	41.9643	1836	3 370 896	42.8486	1911	3 651 921	43.7150	1986	3 944 196	44.5646
1762	3 104 644	41.9762	1837	3 374 569	42.8602	1912	3 655 744	43.7264	1987	3 948 169	44.5758
1763	3 108 169	41.9881	1838	3 378 244	42.8719	1913	3 659 569	43.7379	1988	3 952 144	44.5870
1764	3 111 696	42.0000	1839	3 381 921	42.8836	1914	3 663 396	43.7493	1989	3 956 121	44.5982
1765	3 115 225	42.0119	1840	3 385 600	42.8952	1915	3 667 225	43.7607	1990	3 960 100	44.6094
1766	3 118 756	42 0238	1841	3 389 281	42.9069	1916	3 671 056	43.7721	1991	3 964 081	44.6206
1767	3 122 289	42.0357	1842	3 392 964	42.9185	1917	3 674 889	43.7836	1992	3 968 064	44.6318
1768	3 125 824	42.0476	1843	3 396 649	42.9302	1918	3 678 724	43.7950	1993	3 972 049	44.6430
1769	3 129 361	42.0595	1844	3 400 336	42.9418	1919	3 682 561	43.8064	1994	3 976 036	44.6542
1770	3 132 900	42.0714	1845	3 404 025	42.9535	1920	3 686 400	43.8178	1995	3 980 025	44.6654
1771	3 136 441	42.0833	1846	3 407 716	42.9651	1921	3 690 241	43.8292	1996	3 984 016	44.6766
1772	3 139 984	42.0951	1847	3 411 409	42.9767	1922	3 694 084	43.8406	1997	3 988 009	44.6878
1773	3 143 529	42.1070	1848	3 415 104	42.9884	1923	3 697 929	43.8520	1998	3 992 004	44.6990
1774	3 147 076	42.1189	1849	3 418 801	43.0000	1924	3 701 776	43.8634	1999	3 996 001	44.7102
1775	3 150 625	42.1307	1850	3 422 500	43.0116	1925	3 705 625	43.8748	2000	4 000 000	44.7214

INDEX

A, defined, 83
A_L, defined, 93
A_U, defined, 93
AQL, defined, 201
Abbreviations. *See* Symbols
Absolute basis, 20
Acceptance, false, 103, 105
Accidents, frequency of, 186
Accuracy, 11, 19
  defined, 10
Amber glass, sulfur, 71, 73, 81
Analysis
  for carbon, 126, 145
  duplicate, 27
  of experimental data, 63
  of variance, 44, 133, 146, 155. *See*
    Variance
    glass melting, 148–151
    grouping data, 158
    steps in, 166
Applied research, defined, 1
Attitude of individual, 15
Attributes, 4, 171
  comparison for, 171
  method of, defined, 2, 17
  range for, 173
  sample size, 172
Average, 2
  loss of control, 99
  quality level, defined, 201
  range of, 88, 90
  sample size for, 115–117
  uncertainty of, 86, 91

Batch, as a variable, 42
Batches, using different raw materials, 58
Bias, 25
  conclusions on, 28
  in conclusions, 27
  psychological, 21, 25, 26
  in reading scales, 28
Binomial function, 199
Block
  allocation of variables, 38
  defined, 35, 37
  incomplete, defined, 35
  random order, 38
Block design, defined, 35
Bottles, strength of, 71, 74
Brick strength, cell plotting, 71, 75, 83

c, defined, 4, 173, 192
CR, defined, 197
χ^2, defined, 181
Carbon analysis, 126, 145
Cause, control chart, 33
Cement variables, 43
Chi-square test, 181–186
Classical design, 36, 47, 48, 51
Coding of data, 79, 135, 152
Coin, tossing a, 185
Common sense, defined, 10
  experimentation, 10
Comparative observations, 20
Comparing data, 109
Concept, defined, 9
Conclusion, defined, 10

Conclusions on bias, 28
Conditions of experiment, 14, 15
Confidence. *See* Probability
  envelope, and sample size, 96–98
    for sigma, 96
  interval, 198
  limits of average, 86, 91, 128, 176
  limits of sigma, 94
  sigma from range, 107
Constants, of experiments, 15
Construction, of operating characteristic
  curves, 190
Consumer risk, 197
Continuity, correction, 181, 185
Continuous sampling plan, 205
Control chart, 32
  technique, 16
  trends, 33
Controlled experiment, 14
Controls, use of, 16
Correction for continuity, 181

d, 116, 117, 207
d_1, d_2, 207
df, 88, 158
Data
  analysis of, 63
  defined, 9
  experimental, 19
  limitation of, 1
  periodicity, 32
  presentation of, 23
  required amount for significant differ-
    ences, 113, 115, 172
  trends, 32
Decision
  acceptable risk of, 11
  human, 11
  technical, defined, 9
Defect, 4
  variable, 4
Defectives, 4, 172, 173, 177, 179, 180. *See*
    Attributes estimate of, 211
Degrees of freedom, 88, 120, 131, 181, 188
  for analysis of variance, 158
  for χ^2 test, 188
  for F test, 137
  for pairs of values, 132
Density of glass, 103, 119, 122, 124
Designs
  batch and equipment as variables, 43
  choice of order of experiments, 59
  classical, 36, 47, 48, 51
  experimental, 36, 38, 50, 51
  of experiments, 35
  factorial, 35, 47, 48, 51, 55, 56, 59

  incomplete, 54, 60, 61
  Latin square, 36, 45, 46, 47, 50, 51, 52,
    53, 57
  reducing the experimental work, 54
Development, defined, 1
Dirty statistics, 105
Double sampling plan, 203
Drift
  in apparatus, 36
  machine, 28
Duplicate
  analyses, 27
  runs, 40
    time between, 41
Durability, glass, 129

E, defined, 129, 181
Engineering, defined, 1
  operating, 1
  production, 1
Equipment
  as a variable, 42
  multiple, 42
Error, 9, 11, 21, 135
  common to all experiments, 12
  evaluation of, 16
  estimate of, 38
  independently evaluated, 39
  offsetting, 57
  seeking cause of, 136
Event, defined, 4
Examples. *See* Lists of Examples, Figures,
    and Tables, in front of book
Experience, 13
Experimentation, fundamentals of, 13
Experiments
  controlled, 14
  random order, 31
  types of, 15
Extremes of conditions, 15

f, 83
F, defined, 133, 183
F test, 133, 138–141, 146
  cases for, 137
  critical, 144
Facility and output, 152
Fact, defined, 9
Factorial design, 35, 47, 48, 51, 55, 56, 59
Factorial plan, 47, 48, 51, 55, 56, 59
Freedom. *See* Degrees of freedom
Frequency
  curve, normal, 65
  expected, 181
  observed, 181
  relative, 4

G, 138
Glass
  density, 103, 119, 122, 124
  durability, 129
  iron oxide in glass sand, 85
  melting example, 146
  sodium oxide in, 85
  softening point by groups of data, 87
    by pairs of data, 22
  strength of, 27, 71, 74
  strength of rod, 103, 109, 112, 113, 124, 142
  sulfide sulfur in amber, 71, 73, 81
  tubing manufacture, 71, 72, 103
Graphical comparison of data, 109
Graphical determination of average and sigma, 68, 69, 70
Grouping data, 19, 23
  for analysis of variance, 158
  by cells, 71, 75, 83
Groups, selection of data, 26

h_1, h_2, defined, 207
Hunch, defined, 9
Hypothesis, defined, 10

Idea, defined, 9
Individual, as a variable, 14
Incomplete block design, balanced, 36
Incomplete experimental blocks, 54, 60, 61
Inspection
  device, sigma of, 103, 104
  quality from OCC, 192
  plans, based on OCC, 192
Interaction
  defined, 36, 154
  detection of, 154, 157, 163
Iron oxide in glass sand, 85

Judgment, 11
Jump of data, 32

k, number of values for σ to estimate σ', 3
K
  and confidence level, 93
  defined, 87–94
  factor, 87–93

LCL, 88
LTFD, defined, 203
Latin square
  design, 36, 45–47, 49–53, 57
  limitations, 45
Law, defined, 10
Lead content, in raw material, 130
Limits, acceptance, 27

Literary Digest poll, 27
Logic, 11
Lot tolerance fraction defective, 203

m, defined, 83, 116, 117, 138
  in sequential sampling, 207
m_0, 116, 117
MS, defined, 146, 148–154
Mean square, 148–154
Meaning of data, 63
Meaning of results, 63
Micrometer, use of, 20
Microstatistics, 107

n, defined, 2, 4, 138, 209
N, defined, 2
np, defined, 4, 5, 177–181
Normal distribution curve, 64, 65
Normal distribution law, 32

O, defined, 181
OCC
  basis for inspection, 192
  defined, 190
  interpretation, 192
  for quality range, 194
  and quality, 192
Observation, 2
Observed value, 2
Operating characteristic curves, 190–205
Operating engineering, defined, 1
Opinion, 25
Order, 1
  drawn from a hat, 37
  of experiments, 31, 37
  random, 35
Outline of book, 5

p, defined, 4, 5, 172
P, defined, 64, 89, 118, 190
pn. See np.
PR, producer's risk, defined, 201
Package identification, bias in, 26
Pairs of values, 21, 129, 131
Planning of experiments, 35
Plastic pressing, time, 27
Poisson function, 189, 191
  summation of, 210
Population
  defined, 2
  measured, 88
Precision, 19, 133, 142
  comparison of two sets of data, 133
  defined, 9
  differences in, F test, 137
  from pairs of values, 22

Presentation of data, 23
Principle
    defined, 10
    of randomization, 31
Prediction, 11
Probability
    defined, 64, 89, 118, 190
    and percent defects, 192
    in OCC, 192
    from sigma, 64, 65
Probability paper, 66–70
Producer risk, 201
Production, 1
Production engineering, 1

Q, defined, 213
Quality, 189
    control, 88, 189
        variation from central control line, 33
    evaluation, 189
        sample size for attributes, 213
    level, 213
    range
        for attributes, 212
        from OCC, 194
Quick statistics, 107

r, defined, 198
R, defined, 105, 198
rms deviation, 77
RQL, defined, 197
Random
    order, 35
    sequence, 32
Randomization, 31
Randomness
    accomplishments, 33
    attainment of, 33
    definition, 32
Range
    for attributes, 173
    of average, 105
    effect of sample size, 177
    for estimating sigma, 106
    of quality, OCC, 194
    of $\bar{X}'$ and confidence level, 93
Raw materials, 59, 187
Reasoning, inductive and deductive, 10
Recommendation, 10
Reducing the experimental trials, 54
Rejectable quality level, 197
Rejection
    of data, 29
    false, 103, 105
Replication, 37, 40
    defined, 5, 35

purposes served, 41
required, 11
Research, defined, 1
    fundamental, 1
Residual, in variance, 148, 154, 156, 165
Residual SS, the error, 136, 163
Results
    defined, 9, 16
    negative, 29
    not independent of time, 37
    pairs of, 21
    types of, 16
    unusual, 29
Root-mean-square deviation, 77

σ, defined, 3, 64, 77
σ', defined, 3, 64, 66
σ_D, defined, 119, 128
σs, defined, 128
s, defined, 4, 207
SS, defined, 23, 77, 134, 148, 182
    residual, 136
Σ, defined, 22, 77, 182
Sample, defined, 2
Sample average, $\bar{X}$, 2
Sample fraction defective, 5
Sample plan, double sample, 203
Sample size, 113, 115, 169, 173
    for attributes, 172, 212
    and average quality, 196
    defined, 2
    effect, 66, 115
    and inspection for quality, 196
    and limits of $\bar{X}'$ and σ', 66
    and lot size, 201
    and OCC, 203
    quality evaluation, 209
    and quality range, 194
    for range, 177
    for sequential plan, 209
    for similar CR, 203
    for similar PR, 203
    for variance, 169
Sampling
    comparisons, 201
    continuous, 205
    plans, 189
        sequential, 206
    random, 32
Scale, misreading a, 22
Shift
    and male-female crew, 135, 155
    and output, 152
Sigma. *See* σ. *See also* Contents and Lists
    of Examples, Figures and Tables, in
    front of book

Single measurement, 19
Single value, 20
Significant differences, 109, 113, 172, 173
Significant figures, 23
Skew, of sigma, 95
Sodium oxide in glass, 85
Squares and square roots, 221
Standard deviation. *See* σ *and* Sigma
Standards, defined, 16
Statistical analysis, 10
Statistical control, 32
Subgroups, defined, 2
Sulfide sulfur in glass, 71, 73, 81
Sum of squares, 77
Summation of the Poisson, 210
Symbols. *See* letter symbols in appropriate
    alphabetical order in Index

t, defined, 88, 118, 121
t values, uncertainty of average, 93
Technical community, defined, 1
Technical steps, 1
Telephones, groups with and without, 27
Terminology, 2
Theorem, defined, 10
Theory, defined, 10
Time measurement interval, 92, 123
    sigma for, 93, 123
Transposition of figures, 22
Trends, 32

Trial, defined, 4
Tubing manufacture, glass, 71, 72, 103

u, defined, 4, 5
UCL, 88
Unit, defined, 2
Universe, defined, 2

Value, 11
Variables, 14
    measured, 15
    method of, 2, 17
    number of, 15
    random order of, 33
    uncontrolled, 15
Variance. *See* Analysis of variance *and*
    Variance ratio
Variance ratio, the *F* test, 136, 138
Vitamins, and colds, 26

W, defined, 85
Wear of a part, losses from, 99, 102
Weighted average, 3

x, defined, 83
X_1, defined, 2
$\bar{X}$, defined, 2
$\bar{X}'$, defined, 2, 66, 93
$\bar{\bar{X}}$, defined, 3

Z, defined, 172